PROTOPLASMATOLOGIA

HANDBUCH DER PROTOPLASMAFORSCHUNG

HERAUSGEGEBEN VON

L. V. HEILBRUNN UND F. WEBER
PHILADELPHIA GRAZ

MITHERAUSGEBER

W. H. ARISZ-GRONINGEN · H. BAUER-WILHELMSHAVEN · J. BRACHET-BRUXELLES · H. G. CALLAN-ST. ANDREWS · R. COLLANDER-HELSINKI · K. DAN-TOKYO · E. FAURÉ-FREMIET-PARIS · A. FREY-WYSSLING-ZÜRICH · L. GEITLER-WIEN · K. HÖFLER-WIEN · M. H. JACOBS-PHILADELPHIA · D. MAZIA-BERKELEY · A. MONROY-PALERMO · J. RUNNSTRÖM-STOCKHOLM · W. J. SCHMIDT-GIESSEN · S. STRUGGER-MÜNSTER

BAND VIII

PHYSIOLOGIE DES PROTOPLASMAS

3

MOTILITÄT

a

PROTOPLASMIC STREAMING

WIEN
SPRINGER-VERLAG
1959

PROTOPLASMIC STREAMING

BY

NOBURÔ KAMIYA
DEPARTMENT OF BIOLOGY, FACULTY OF SCIENCE,
OSAKA UNIVERSITY

WITH 82 FIGURES

WIEN
SPRINGER-VERLAG
1959

Softcover reprint of the hardcover 1st edition 1959

ISBN 978-3-211-80524-4 ISBN 978-3-7091-5750-3 (eBook)
DOI 10.1007/ 978-3-7091-5750-3

Protoplasmatologia
VIII. Physiologie des Protoplasmas
3. Motilität
a) Protoplasmic Streaming

Protoplasmic Streaming

By

NOBURÔ KAMIYA

Department of Biology, Faculty of Science, Osaka University

With 82 Figures

Contents

I. Introduction

Since CORTI (1774) observed for the first time a streaming in plant cells, a history of more than 180 years has already elapsed in the study of protoplasmic streaming. During this period of time, many experiments have been conducted and various discussions have been made on protoplasmic streaming, each of them within the limits of the knowledge and techniques of its own times and more or less under the influence of the ideas that prevailed then. With such a long history of research, in spite of all the interest and efforts that have been devoted to it by many, the mechanism of this intriguing and complicated biological phenomenon is still left far from clear.

It is an undeniable fact that one important factor stood in the way in analysing this phenomenon, that is, the limitations in the techniques applicable to this research. Another factor lay in the fact that we still have not enough knowledge about the protoplasm regarding its structural organisation, mechanochemistry, energetics, rheology, etc., all of which constitute basic problems in understanding the nature of the streaming.

A favourable point about the protoplasm in observing its streaming is that the protoplasm is usually replete with numerous granules of various sizes and kinds. Were there no such granules which serve as indices, it would certainly be difficult to recognize the flow even though protoplasm is streaming. On the other hand, the results obtained with these granules as indices of flow entail the question whether or not the visible movement of granules shows the flow of the ground plasm (hyaloplasm) in which they are embedded, or whether it is only the granules themselves that actively move. The opinions of the authors in the past on this point are not always unanimous, yet ample evidences show that there are many cases in which granules are carried along passively with the matrix. As a rule, we shall here deal with the bodily flow of the ground plasm forming the matrix, but in order to discuss it, we shall necessarily refer to the free movement of the granules, since, as will be considered in a later chapter (p. 171), the "independent" or "autonomous" movement of the granules and the streaming of the matrix are phenomena so closely related that it is scarcely possible to consider them separately.

Previously, not a few workers regarded the streaming as an abnormal phenomenon occurring when the cell is injured or when it is subjected to environmental changes (FRANK 1872, DEHNECKE 1886, KELLER 1890, KRETZSCHMAR 1904, etc.). HAUPTFLEISCH (1892) distinguished the protoplasmic streaming in two groups, the *primary* and *secondary*. The former is the streaming observed in a cell under normal and natural conditions while the latter is the streaming which is established anew through physical or chemical stimulus in a cell whose protoplasm has been in a quiescent state. VELTEN (1872) thought that the protoplasmic streaming is the most common phenomenon appearing at a certain stage of life of protoplasm in a cell. Further, SEIFRIZ (1943, 1952) regarded protoplasmic streaming as a general basic phenomenon occurring in a cell with active metabolism. The protoplasmic streaming in wholly intact cells has been reported by many authors (SCHORR 1935, LINDENBEIN 1937, DOI 1950, etc.). We must also take into account the possibility that very slow and therefore hardly recognizable streaming may be carried on in many kinds of cells even though their contents appear to be in rest. Such a slow motion can be visualized impressively by the aid of time lapse motion picture. Though we naturally admit that there are cases in which protoplasmic streaming is induced by a wound, we know also that there are many "primary" streamings in nature that are not related to wounds. Probably we are right to suppose that protoplasmic motion is not totally absent in a metabolically active cell on a submicroscopic level. Yet for the development to a visible motion to occur, there must be a certain orderliness in respect to the phase and direction of the motor mechanism.

The vast literature of work dealing with the protoplasmic streaming since the beginning of its history up to around 1930 was listed by Yamaha and Ida (1933). When we look at the exhaustive descriptions of the protoplasmic movement in Hofmeister's classic entitled "Die Lehre von der Pflanzenzelle" published as early as 1867, we are impressed by the fact that basic knowledge about protoplasmic streaming had already been so well established by that time. Among the treatises or review articles concerning more or less general problems in this field there is the classical, celebrated book of Ewart (1903) in which knowledge about the protoplasmic streaming up to his time is well summarized into a unified whole together with the results obtained by his own experiments. General articles dealing with the streaming of protoplasm by Lundegårdh (1922) and passages in Küster's "Pflanzenzelle" (1935, 1951, 1956) are also valuable key references on past activities.

Since two decades or more new activity has been brought about in the study of protoplasmic flow through the introduction of slime mould culture into laboratories. It was Seifriz who hit upon *Physarum polycephalum* and played a prominent part in leading recent progress and popularization in the study of protoplasmic streaming through this material. Seifriz's review article (1943) on protoplasmic streaming with its broad scope and with its coverage of both old and new literature and his later article (1952) are also valuable sources of knowledge in this field including the study of the slime mould [1].

In the following more space was devoted to the papers published recently as well as to the experiences and observations of the author himself and his associates. A partial excuse for this may be found in the existence of the above review articles and general treatises.

II. Types of Streaming

1. General survey

The state and mode of protoplasmic streaming are diverse. For convenience' sake, they may be classified in several patterns which we shall describe below.

[1] There are also some general articles dealing with protoplasmic streaming published in the author's native language (Ohta 1950, Kamiya 1953, 1957).

a) Agitation

What was called *agitation* (Seifriz 1943) "for want of a better word," or later *turbulent motion* by the same author (Seifriz 1952) "for want of another word," is regarded as the most poorly organized type of streaming on the microscopical level, although it is most common and prevalent. This type of movement probably corresponds to the motion designated as "Glitschbewegung" long before by Nägeli (1855, 1860) and as "Digressionsbewegung" by Wigand (1885) (cf. also Hauptfleisch 1892, p. 175, Lundegårdh 1922, p. 364). Nägeli (1855, p. 49) observed, already more than a century ago, what he called "Glitschbewegung" of particles gliding on the interface between cytoplasm and cell sap in many *Conjugatae* (especially *Closterium, Penium, Spirogyra,* etc.), and also in *Achlya* etc. They are partly at a standstill, partly gliding toward the end, and partly toward the center of the cell. The motion is erratic and haphazard, yet it is, statistically speaking, not totally devoid of directional control such as is the case in Brownian motion.

The opinion was presented in the literature (Meyer 1920) that Brownian motion and protoplasmic streaming might be related or might have the same cause. As a matter of fact the alteration of the motion of a particle is performed in such a way that no sharp distinction is possible. Actually, however, agitation (Glitschbewegung) is distinct from Brownian motion. While Brownian motion is a phenomenon caused by thermal agitation of molecules, the so-called agitation of protoplasm is a motion which is brought forth by consuming metabolic energy. In agitation, small particles slide smoothly without turning aside over a stretch which is larger than that of Brownian motion. Inconspicuous though this type of protoplasmic motion may appear, it is significant as a primitive type of movement from which more orderly streaming develops and also as showing a wide occurrence in many active cells.

Recently Jarosch (1956 c) presented a reasonable and yet practicable method to express the degree of regularity and intensity of this motion quantitatively. As is well known, Einstein's formula for Brownian movement may be expressed as follows:

$$\overline{\Delta X^2} = \frac{k\,T\,i}{3\,\eta\,r}$$

where $\overline{\Delta X^2}$ is the mean square displacement of a particle in a time interval i, T is the absolute temperature, η the viscosity, r the radius of the particle, and k a constant. The formula thus shows proportionality between ΔX^2 and i. Hence, if the time interval (i) is made twice as long, namely, $2\,i$ is substituted for i, the mean square displacement will also be twice as large, or $2\,\overline{\Delta X^2}$. This proportionality is, however, valid only for an utterly irregular motion as is the case of Brownian movement of a particle.

If the time interval is made twice as long for the movement having a perfect orientation and steady rate, it is clear that the mean square displacement ($\overline{\Delta X^2}$) must be 4 times as large. The degree of regularity of all the intermediary motions between the two extremities, namely, the motion of complete randomness and the motion of constant orientation and speed, can be expressed by a value between 2.00 and 4.00. This value representing the ratio of the mean square displacement of double intervals to the mean square displacement of single intervals, can serve as an objective indication of the degree of regularity in the motion of a particle. JAROSCH called it the regularity quotient (*Regelmäßigkeitsquotient*) or *Qr*. According to the above reason *Qr* value increases from 2.00 up to 4.00 with an increase in regularity in respect to both orientation and speed. A practical application of this method to the protoplasmic motion in the cell of the inner epidermis of *Allium cepa* will be referred to later (p. 16).

b) Circulation

The streaming designated as *circulation,* which is characteristic of the cell having transvacuolar protoplasmic strands, is easily recognized in large cells of plant hairs, such as stinging hairs of *Urtica,* hair cells of *Cucurbita, Gloxinia, Tradescantia, Saxifraga, Campanula* (stylus), etc.; further, in *Spiroyra* and other *Conjugatae,* in cells of berries (*e. g. Symphoricarpus racemosa*), in parenchymatous cells of many monocotyledons (*e. g. Allium cepa*), etc. HOFMEISTER (1867), VELTEN (1872), BUSHEE (1908), LUNDEGÅRDH (1922), etc. reported many materials suitable for observing protoplasmic streaming of various types including circulation.

The behaviour of granules in a state of circulation, which are found in or on the protoplasmic layers adhering to the cell wall or strands crossing the vacuole, has been described repeatedly (cf. KÜSTER 1933 b, 1956). The starch grains in the colourless rhizoids of *Caulerpa* can be many times as thick as the protoplasmic strands which carry them. When two such grains, which have been sliding along a strand in opposite directions, encounter each other, either one of them is drawn back, or one makes way for the other (JANSE 1890). In a state of circulation direction and speed of granules found in or on the protoplasmic layers on the walls or in or on strands connecting them across the vacuole are still haphazard and unstable KÜSTER (1933) described the circulation in the cell of the stamen hair of *Tradescantia.* Microsomes of various sizes in close proximity differ in the speed and direction of their movement and give the impression that there are very complicated courses of streaming in the cytoplasm. When two particles come from opposite directions and meet, or when one particle comes from behind and overtakes another, these particles continue their movement each with its own speed even after they subsequently fall apart from one another.

BÜNNING (1935), who investigated the streaming of the particles with a special condenser enabling color contrast illumination (Micropolychromar, Zeiss), said that each particle moves with its own independent speed and that the hyaloplasm of protoplasm is a medium which is quiescent and has elasticity. This idea was expressed a long time ago by NÄGELI and

SCHWENDENER (1867). We shall come to this problem later again (p. 71). BÜNNING is further of the opinion that the particles move not within the protoplasmic thread, but along its surface.

In the perfectly intact onion cell, SCHORR (1935) followed with the aid of episcopic illumination (Ultropak, Leitz) the course of the movement of a small anthocyan crystal attached to the protoplasmic layer lying on the cell wall. Fig. 1 is a record during 64 minutes, the numbers showing in sequence the positions of an anthocyan crystal.

In the case of circulation, transvacuolar strands are subjected to continuous change in thickness, form and location; a strand may be shifted aside, branched, united, may retract, or may emerge *de novo* from the layer of protoplasm lining the wall. The streaming courses of the protoplasm in the layer adhering to the wall are also subjected to constant alteration. Strictly speaking, therefore, the same circulation pattern never appears twice throughout the life of a cell. According to KÜSTER one of the most important points distinguishing circulation from rotation is that the former is periodic while the latter is aperiodic. HOFMEISTER (1867, p. 38) described the reversal of the streaming direction in the protoplasmic strand of the *Tradescantia* hair as follows: "Die Richtung der Bewegung bleibt in keiner Strombahn dauernd dieselbe. Sie setzt nach einiger Zeit in die entgegengesetzte um. Aber wie bei *Tradescantia* die Gestaltänderungen des Netzwerkes aus Strängen strömenden Protoplasmas langsamer sind als bei den Myxomyceten, so sind auch die Perioden um vieles länger, in denen die Stromrichtung wechselt. Sie bleibt in derselben Bahn 10 bis 15 Minuten konstant, verlangsamt sich dann sehr plötzlich, stockt während einer veränderlichen, zwischen wenigen Sekunden und mehreren Minuten schwankenden Frist, um dann in die entgegengesetzte überzugehen. Die neue Stromrichtung tritt langsam ein, wird aber sehr rasch beschleunigt. Auch bei *Tradescantia* ist es vollkommen deutlich, daß die neue Bewegung nach rückwärts sich fortpflanzt: es werden fortschreitend Teile des momentan zur Ruhe gelangten Protoplasma in die Strömung hineingezogen, welche dem Zielpunkte sukzessiv ferner liegen."

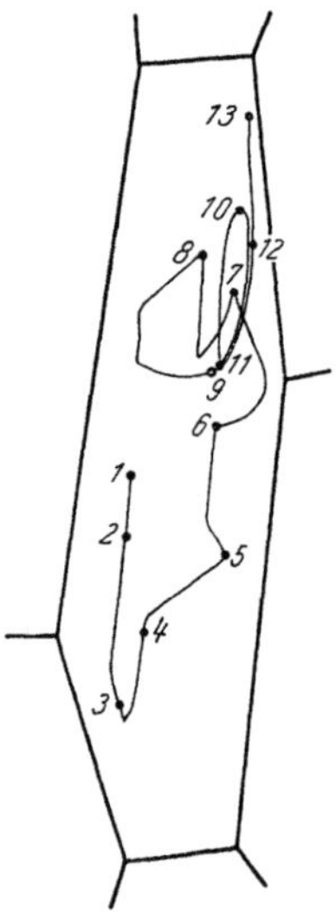

Fig. 1. Wandering of an anthocyan crystal body in an epidermal cell of *Allium cepa* (observation of uninjured material through episcopic illumination). Intervals in minutes between adjacent two points designated by serial numbers are in parentheses as follows: 1 - (3) - 2 - (2) - 3 - (6) - 4 - (9) - 5 - (4) - 6 - (13) - 7 - (8) - 8 - (7) - 9 - (6) - 10 - (4) - 11 - (11) - 12 - (1) - 13. (SCHORR 1935.)

Indeed, it is "periodic" in the sense that the direction of streaming changes alternately in circulation, but the interval of the streaming in one direction or the other is hardly constant.

c) Rotation

If protoplasm is limited to the periphery of a cell, and if protoplasm streams like a rotating belt, we speak of that streaming as *rota-*

tion[2]. In contrast to circulation, the configuration of protoplasm remains nearly constant as the streaming is more or less stationary in this case. Rotation is thus the most orderly of various types of protoplasmic streaming and serves as an appropriate object of quantitative study.

This type of motion is often found in leaf cells of aquatic plants (*e. g., Elodea, Vallisneria, Sagittaria,* etc.), root hair cells and pollen tubes of many plants, cambium cells (*e. g., Sida Napaea, Lathyrus heterophyllus, Artemisia absinthium, Sambucus nigra,* species of *Tradescantia,* etc.) and especially in cells of *Nitella* and *Chara.*

Since CORTI (1774), TREVIRANUS (1806) and AGARDH (1826), the streaming in *Characeae* has been studied very often. It is distinguished from the streaming in *Elodea* or in many other plant cells in that the majority of chloroplasts are not carried about throughout the cell but settle down in the resting protoplasmic layer on the wall, as long as the normal order of the cell is kept. The direction of streaming in cells of *Chara* or *Nitella* is fixed regularly. For instance, in the leaf the upward streaming is found on the outer side of the cell, while the downward streaming is on the inner side of it (cf. AGARDH 1826, p. 127).

When upward and downward streamings pass by each other there are two distinct zones having no or few chloroplasts, the so-called indifferent lines. It was pointed out by VOTAVA (1914) in *Nitella flexilis* that the cell membrane develops an edge of inconstant width projecting slightly inward at the indifferent line. Chemically, these zones (in the rhizoid and the colorless internode of *Chara*) are distinguished from the other parts in that they are blackened by 1% silver nitrate (LINSBAUER 1929). But BARG (1943), who worked on leaf cells of *Chara foetida,* could confirm neither VOTAVA's observation nor the blackening with silver nitrate reported by LINSBAUER. Chloroplasts lying near this zone usually take a rounded form instead of an elongated one (JAROSCH 1956 b), although the case is also known in which chloroplasts near or on this zone are characterized by long-stretched or hook-like form (BARG 1943). Protoplasm flows along the spiral course the inclination of which coincides with that of the indifferent zones and that of chloroplast alignments. The inclination of the spiral changes according to the stage of growth of the internodal cell (GREEN 1954, GREEN and CHAPMAN 1955). Among the old literature on the protoplasmic streaming in the cell of *Characeae,* HÖRMANN's extensive work (1898) is especially worthy of note. We shall consider the streaming in *Characeae* more in detail in later chapters.

---

[2] KUSUNOKI (1943) proposed a new terminology in German *Umlaufbewegung* for this type of protoplasmic streaming in place of *rotation* as the latter word is likely to be misunderstood as meaning revolution of an object on its own axis, when actually protoplasm moves around along the cell wall. Though the author fully understands the logical correctness of KUSUNOKI's proposal he used the word rotation here partly because of its traditional usage and partly because a proper English expression could not be found for *Umlaufbewegung.* Protoplasmic streaming in plant cells having cell walls is often referred to as "cyclosis." This terminology, in which are usually included both rotation and circulation, is familiar to zoologists.

d) Fountain streaming

There is an intermediate type between circulation and rotation. It is the "fountain-like rotation" (*springbrunnenartige Rotation,* cf. BERTHOLD 1886, p. 119; LUNDEGÅRDH 1922, p. 367), in which a thick central strand of protoplasm moves acropetally or basipetally while the protoplasmic layer on the wall moves in the opposite direction. This type of motion is to be seen in root hairs (*Trianea bogotensis, Hydrocharis Morsus ranae,* etc.), pollen tubes of many plants, etc. The young hairs of *Trianea* etc. show a typical circulation with many, actively moving and ever changing strands, which occasionally fuse into a central strand presenting the appearance of a fountain. Generally, the flow from the base to the tip occurs in the layer held against the wall and that from the tip to the base takes place along the central strand which penetrates the vacuolar space longitudinally (reverse fountain type). When a pollen tube elongates to a certain extent, the rotation in the pollen grain finds a way into the pollen tube. The streaming is of a typical fountain type in *Plantago major, Lilium Makino,* and of a reverse-fountain type in *Lochnera rosea, Camellia japonica* (KUMAGAYA 1950).

Besides the patterns of streaming described in the foregoing there are still different types of flow throughout the plant kingdom which need special attention.

e) Protoplasmic streaming occurring along definite tracks

In the protoplasmic layer lining the cell wall there are often linear tracks along which protoplasm and chloroplasts stream. They constitute a sort of striations on the cortical gel and are separated from each other with zones where protoplasm remains quiescent.

Beautiful examples of the streamlets along the tracks are found in some coenocytic algae. In *Acetabularia,* the streamlets are found along tracks of indefinite number arranged parallel to the longitudinal axis of the cell. The direction of the streamlets being different according to the tracks, several streams are seen to go in opposite directions simultaneously in one and the same cell. The characteristic feature of the streaming of chloroplasts is observed in the stalk, hair-like "shoots" and vegetative whorled "leaves" in the stage before the cap is accomplished. The tracks on which streaming takes place are to be recognized especially clearly as striations if the cell is centrifuged so that all the chloroplasts are collected at one end of the cell. When chloroplasts are pushed aside, what is seen behind are many striations running in parallel with unequal intervals on the cortical layer (Fig. 2). The protoplasm forced to one end of the cell together with chloroplasts always comes back along these tracks. The width of the track in a stalk of *Acetabularia calyculus* is about 2–3 μ which amounts to only $^1/_3$–$^1/_5$ the width of a chloroplast (TAKATA 1958).

According to KÜSTER (1933 c) the streamlets in palisade tubes of *Codium bursa* usually have the width of one, rarely 2–5 chloroplasts. The tracks of the streamlets run straight in the longitudinal direction or spirally.

The same is also observed occasionally in *Bryopsis* tubes having scanty chloroplasts. Protoplasmic streaming in *Caulerpa* presents one of the most characteristic features as described by JANSE (1890), DOSTAL (1929), etc. JANSE (1890) showed how the track of streams is influenced by external agents. If a small cut is made transversally on a *Caulerpa* leaf, we see that the stream is bent aside shortly above the lesion and at the two ends of the injury the stream is directed linearly to the petiole.

In the sporangiophore of *Phycomyces*, straight tracks, which are close to each other, develop running longitudinally. They are easily distinguished from the hyaline surroundings through their rich content of granules. In old hyphae streamlets leap forward into the vacuole space as fine strands

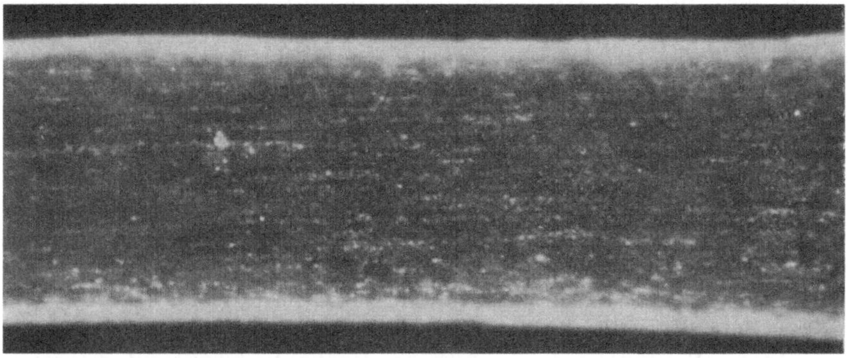

Fig. 2. The stalk of *Acetabularia calyculus* after the chloroplasts and most of the endoplasm are pushed down by centrifugation, leaving parallel striations behind. Width of the stalk: 300 μ (lateral illumination). (Original.)

(KIRCHHEIMER 1933). Streamlets often show spiral courses, the direction of torsion of the striations usually corresponding to that of the membrane of the sporangiophore.

KOPETZKY-RECHTPERG (1938, 1949) described a characteristic arrangement of longitudinal strands in the external, non-moving cortical protoplasm in a desmid, *Netrium digitis*. It is along the channels between these strands that protoplasmic streaming takes place.

f) Protoplasmic streaming in fungal hyphae (tidal streaming)

Protoplasmic flow occurring in coenocytic hyphae of *Phycomycetes* (*Mucor, Rhizopus, Phycomyces,* etc.), which is often referred to as tidal streaming, shows unique aspects. Protoplasm and cell sap are driven in hyphae either in acropetal or basipetal direction. This streaming which has been known since WORONIN (1866) and has been thought to be brought about through a turgor gradient was discussed especially by ARTHUR (1897), TERNETZ (1900), SCHRÖTER (1905), ANDREWS (1912), GREHN (1932), JAHN (1934), etc.; the contents of the cell flow in one direction or the other as water is removed by osmotically active substances or by transpiration either at the apical or basal parts of hyphae. ANDREWS (1912) described only an acropetal streaming.

The streaming occurs by snatches toward the growing tip of the hyphae or the protoplasm is carried away to other hyphal systems through anastomosis. After a resting pause a new streaming commences in the opposite direction.

According to Arthur (1897), Schröter (1905), Buller (1933), etc., there are in hyphae of *Mucor* and *Phycomyces* acropetal streaming of the cell sap together with the inner part of the cytoplasm while the outer layer of protoplasm streams toward the base. The protoplasm in the peripheral layer thus forms a moving mantle around the inner flowing protoplasm. This pattern of streaming corresponds to what Seifriz (1943, 1952) designated as "sleeve" type. Arthur (1897) described his observation as follows: "When the return current exists it is generally well defined, and always occupies the periphery of the hyphal cavity. It carries no vacuoles, and can only be detected by the movement of the microsomes. As the hypha is seen in optical section, there appears to be the usual surging, vacuolated stream moving through the centre, and on either side next to the walls a narrow uniform layer of clear protoplasm, in which minute particles are moving in opposite direction to the central stream. Between the two streams is a quiescent partition of protoplasm of about the same thickness as that lining the cell-walls. These quiescent layers vary much in thickness, and are at times so thin as to be scarcely discernible. It is obvious that whenever such return currents exist, their volume is ample to bring back all the protoplasm borne forward by the central stream, especially as it has become freed from all vacuolar sap, which has probably been used in growth or extravasated." The author himself could also confirm the sleeve-type double flow in hyphae of *Phycomyces* and *Rhizopus*. This type of motion is well observable especially when the entire hyphal system is submerged. Thus we see that Andrews' denial of the occurrence of the peripheral reversed flow of protoplasm was not well based. The sleeve type of protoplasmic streaming is also observable in the sporangiophore of *Phycomyces*, where the peripheral endoplasmic layer moves in a tubular stream toward the substratum while the concentric inner layer around the central vacuole moves toward the top (Pop 1938). According to Pop, the rate of the basipetal streaming (3–4,5 μ/sec.) is usually greater than that of the ascending stream of the inner layer (2–3 μ/sec.).

Although protoplasmic streaming in the non-septate hyphae of the *Phycomycetes* is a well-known phenomenon, it is also reported that very active streaming takes place in the septate hyphae of certain *Ascomycetes* as well. In *Ascophanus carneus* (Ternetz 1900) the entire contents of the hyphae streams as a whole in one direction. A quiescent cortical layer is not visible, but the rate is highest at the axis of the hyphae and diminishes from the axis to the periphery. The septum having a central hole appears to display only insignificant or no resistance to the protoplasmic flow. Jahn (1934) made a detailed study on *Humaria*.

It is known that protoplasmic streaming takes place in the mycelium of species belonging to all the great groups of fungi: *Phycomycetes, Ascomycetes,* and *Basidiomycetes*. Buller (1933) states: "In the *Mucorineae* streaming is characterized: (1) by frequent reversals in the direction of flow of the protoplasm; (2) at least sometimes, by a peripheral current of protoplasm flowing in a direction opposite to that of the main stream; and (3) by a general movement of the vacuoles. On the other hand, in the higher fungi streaming is characterized: (1) by the flow of the protoplasm for an indefinitely long time in one direction only, without any more or less rhythmic reversals in direction; (2) by the absence of peripheral currents of protoplasm flowing in a direction opposite to that of the main stream; and (3) by the fact that the flow of protoplasm often takes place through a cell for long periods of time without the vacuoles being involved in the movement."

In a very young stage of development in which the hypha is still filled with granule-rich protoplasm, no streaming is to be observed. Streaming usually becomes recognizable only after the hypha has formed branches and many terminals or branched parts of hyphae have grown out from the hanging drop into the air.

As the cause of pulsation in protoplasm TERNETZ (1900, p. 285) postulated the local variation in turgor pressure in a rich-branched hyphal system. By means of alternate enlarging and reducing of vacuoles in various parts the protoplasm must be driven in either forward or backward direction. The period appears to be extremely irregular.

g) Protoplasmic streaming in myxomycete plasmodia (shuttle streaming)

Myxomycete plasmodia, commonly known as slime moulds, show characteristic features in their protoplasmic streaming in various respects. The rate of flow as well as the amount of protoplasm carried along with the streaming is exceedingly great as compared with the ordinary protoplasmic streaming in plant cells. Moreover, the direction of the streaming alternates according to a rhythmic pattern. The protoplasmic streaming in this organism has a characteristic in common with the amoeboid movement in that the streaming of endoplasm involves changes in the contour or form of the organism and generally also causes locomotion. Therefore this type of movement can also be well classified in the category of amoeboid movement. The author has in this treatise devoted considerable pages to this type of motion not only because it is the most active and the most conspicuous among the streamings of protoplasm, but also because many recent investigations on the protoplasmic flow have been conducted on this material.

Among the classical studies on protoplasmic streaming the wide-scoped experiments by VOUK (1910, 1913) are particularly worthy of note. During the last two decades various studies have been undertaken by SEIFRIZ and many others using *Physarum polycephalum* as an experimental material. *Physarum polycephalum* can be made to grow easily and abundantly on moist filter paper with rolled oats after the method of CAMP (1936). As this plasmodium grows fast, a great quantity of protoplasm can be provided for the experiments by means of a simple culture. Thus it is a material highly favourable for carrying out experimental work on protoplasm. In regard to general morphology, physiology and culture of myxomycetes and its movement, there are reports by HOWARD (1931), JAHN (1932), MANGENOT (1934), CAMP (1937), COHEN (1939, 1941), DALLEUX (1940), MARTIN (1940), LEWIS (1942), SEIFRIZ (1942), SOBELS (1950), ANDERSEN and POLLOCK (1952), JOHNSON and MOOS (1956), etc.

When a small bit of protoplasm is placed on an agar surface or on a coverslip kept in moist air, it spreads in an hour or two into a disc-like sheet. In this sheet of protoplasm channels are formed along which the plasmasol streams very actively. Streaming takes place only in one direction at a time, but the direction reverses alternately at every one minute or so. SEIFRIZ (1943, 1952) referred to this type of motion as "shuttle streaming." Later the disc-like plasmodium spreads out to form a network and then a branched system of protoplasmic channels with the fan-like plasm expanses at their terminals.

Among varieties of plasmodia, we see another type of flow which is quite different from *Physarum* in its mode of protoplasmic streaming. *Reticulomyxa filosa* described by NAUSS (1949) possesses outstanding characteristics; several parallel lines of continuous and more or less constant flow that alternately streams in opposite directions exist in a broad, thin sheet of protoplasm. As many as twelve channels have been observed to occur. The granules carried by the stream in one channel move very close to those in another. There is no visible wall separating them nor any indifferent line such as seen in an internodal cell of *Characeae.* Such wide bands of protoplasm with alternately opposed flow, which, according to NAUSS, were not reported in similar organisms before, are the material which deserves our attention when considering the mechanism of the motion.

2. Change in streaming type

a) Transition of streaming patterns in intact cells

Of various types of protoplasmic streaming there are some which are apparently different from others, but there are also some which are mutually related by intermediate stages. For instance, agitation, circulation, and rotation are mere conventional terms for expressing the states of flow according to the degree of orderliness, with no clearcut distinctions among them.

Moreover, we can observe in one and the same cell the transition from agitation to rotation or *vice versa* as a process occurring under normal cytogenesis or as a process induced artificially. Taking an example from pollen grains (*Lilium auratum*), the first sign of protoplasmic movement in them after being sown is agitation, which, however, gradually transforms itself into rotation *via* circulation, germination occurring only after about 80 minutes (IWANAMI 1952, 1956). Near the tip ("cap-block") of the young pollen tube the movement exhibits a reverse fountain type (Fig. 3).

In the *Elodea* leaf, we can follow in intact cells the gradual realization of rotation from the state of agitation by way of circulation (cf. VELTEN

1872, p. 652; HAUPTFLEISCH 1892, p. 193; FITTING 1925, p. 328). Change in pattern proceeds also in a reverse direction from rotation back to agitation according to conditions.

Transformation of the streaming pattern, agitation ⇌ circulation ⇌ rotation, is a wide-spread phenomenon and is observable in many plants other than *Elodea*. Generally, in the stage of agitation, small currents are produced in great number in many small loci of a protoplasmic system. This means reciprocal shifting among the structural elements of the cytoplasm at many irregular, unstable interfaces. At the level of the cell as

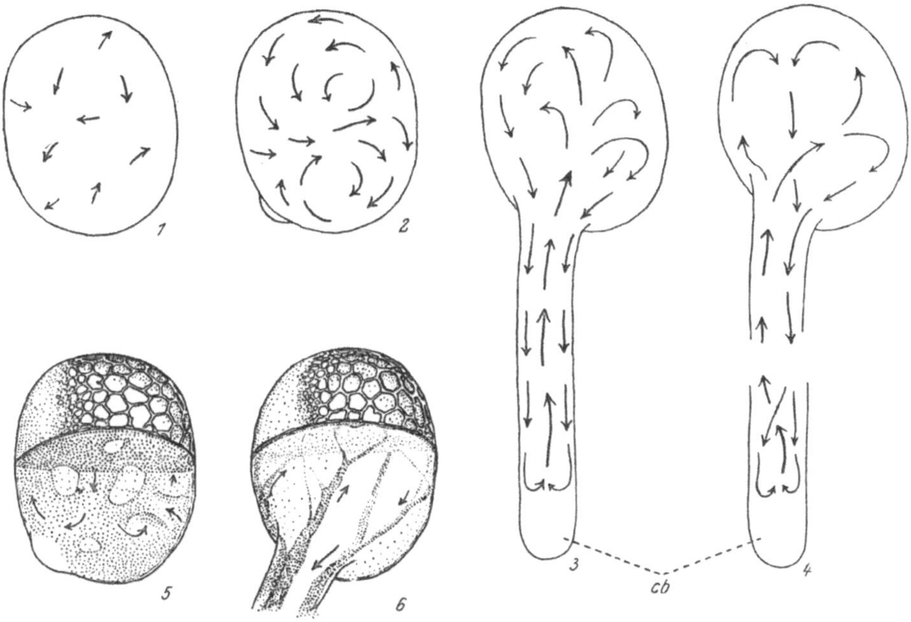

Fig. 3. Pollen grains and young tubes of *Lilium auratum* representing agitation, circulation and rotation of protoplasm. Near the cap-block (*cb*) the movement is always in the manner of a reverse-fountain. (IWANAMI 1956.)

a whole, however, the complete orientation of the elements is not achieved yet, as is shown by the disordered disposition of the direction of the small currents. Agitation, therefore, represents "short-range order" (cf. FREY-WYSSLING 1953, p. 107) in respect to protoplasmic streaming. In course of time, the tendency to orientation further spreads over the entire cell and eventually complete dynamic organization is achieved which renders protoplasmic rotation in a definite direction possible.

JAROSCH (1956 c) made a thorough study on the development of order of the streaming in epidermis cells of *Allium*. In the inner epidermis of onion, protoplasmic motion begins several minutes after preparation. It is interesting to observe the motion when it is initiated. The spherosomes, which have shown agitation or weak Brownian motion at first, start to display spasmodic motion in various directions. These movements are so haphazard, that one cannot tell how and in what direction each particle will move. Particles are often thrown into Brownian movement again after they have travelled over a wide range. Usually, agitation becomes stronger with time and more and more particles participate in the motion eventually bringing forth the state of flow. In a much later stage all particles move equally in almost the same direction. In these "old" streamings, protoplasmic strands are formed across the vacuole emerging from the protoplasmic layer on the wall.

JAROSCH showed the gradual development of orderliness in protoplasmic motion by a method developed by himself (see p. 6). In order to do this, the tracks of movement of spherosomes were first determined by means of an Abbe camera lucida in such a way that the instantaneous position of a granule is marked at every 1.5 sec. Fig. 4 shows these records in one and the same locus of the same cell in different lapse of time after preparation. Here we notice clearly how the state of agitation shifts to more orderly states. JAROSCH's regularity quotients *Qr* for I, II, III und IV in Fig. 4 amount respectively to 2.76, 3.45, 3.73 and 3.81 as inferred from the graph in his paper (JAROSCH 1956 c, Abb. 3).

b) Development of order in an isolated drop of protoplasm

YOTSUYANAGI (1953 a) reported an instructive example of the transition from agitation to circulation up to rotation in an isolated drop of protoplasm from the internodal cell of *Nitella* or *Chara*. He obtained the isolated drops by cutting the cell of *Nitella* or *Chara* and by squeezing out the contents forcefully into artificial media [glucose, KNO_3, $CaCl_2$, $Ca(NO_3)_2$] of unequal hypertonicities.

The protoplasm thus isolated from the cell does not disperse in the environmental milieu in virtue of a new membrane formed immediately at its surface. The protoplasm does not present any type of movement at first, probably due to gelation provoked by the operation, but before long the Brownian movement of fine cytoplasmic granules is resumed, which becomes active with time. Soon after the beginning of Brownian movement in some loci of a protoplasmic drop, the displacement of granules shows a tendency to be so oriented that local streamlets are produced. There we see the transition from the state of disordered Brownian movements to a state of order in small loci though the direction of each of these streamlets remains in disorder. This state of eddy current corresponds to that of the agitation which is occasionally manifested by the cells of *Elodea* leaf or of *Allium* epidermis.

In course of time the streamlets become further unified, progressively forming larger but fewer currents, which are eventually combined to a single rotation either to the right or left. As shown by YOTSUYANAGI's experiment, rotation once established often returns to a less ordered state, then it again passes into rotation. It is at that moment that the reversal of the direction of rotation can take place, a fact which is never found in the normal movement of *Characeae*.

In an *Elodea* cell YOTSUYANAGI also observed a similar behaviour of protoplasm in a small protoplasmic fragment of about 2–3 μ obtained when the cell was thrown into plasmolysis. In those of about 2–3 μ, sizes

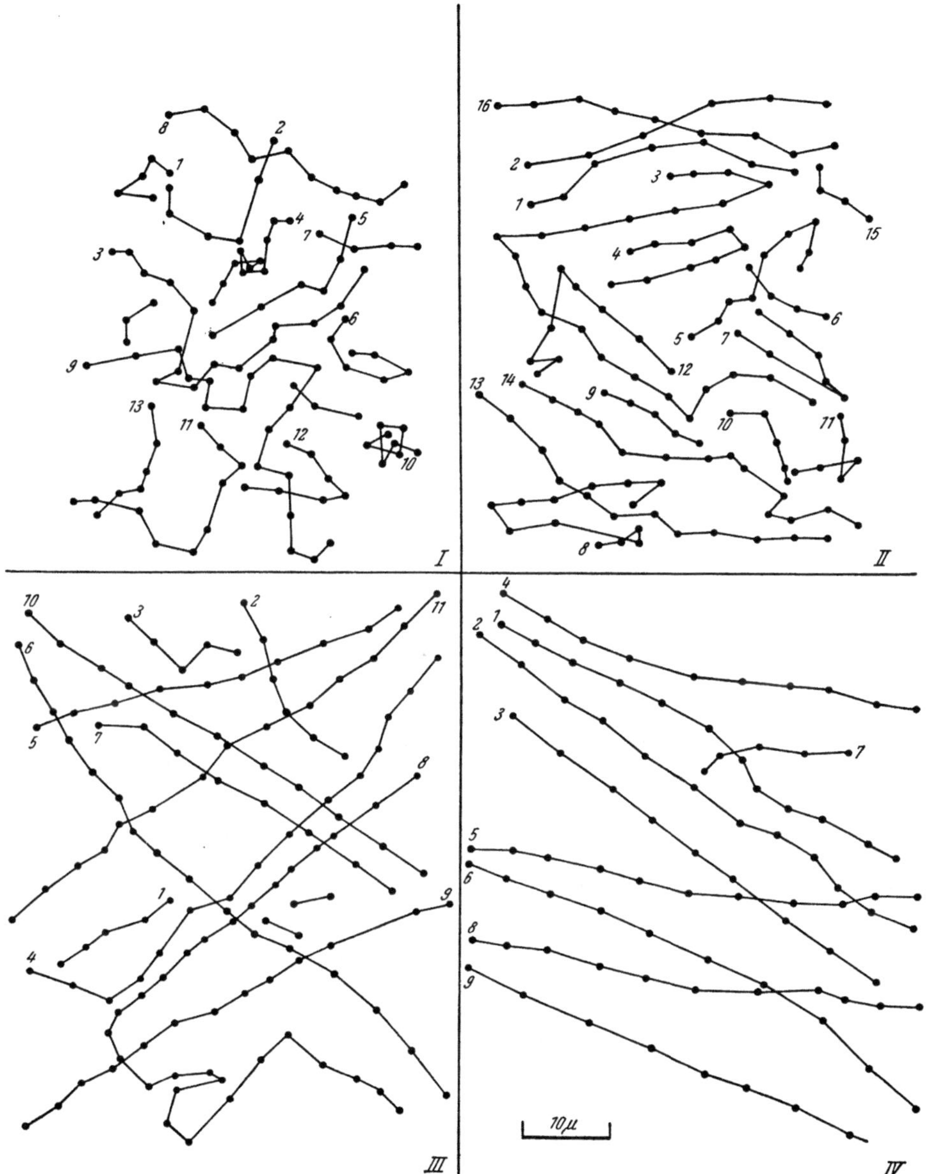

Fig. 4. Courses of moving spherosomes in the protoplasmic layer on the cell wall *(Allium cepa)*. *I*: soon (ca. 5 min.) after preparation, *II*: 40 minutes later, *III*: 90 minutes later, *IV*: next day. (JAROSCH 1956 c.)

which approximately correspond to the dimension of each of the small elementary currents formed in the protoplasm in agitation, rotation may occur, whereas only agitation or circulation is to be seen in the main plasmolysed protoplast of *Elodea*. This is perhaps due to the situation that in a small droplet of cytoplasm an oriented arrangement of the fibrous microscopic elements is already sufficiently established while a wider-range order throughout the larger protoplast is not realized yet with the result that agitation or circulation still reigns in the larger protoplast. Thus isolated protoplasm shows us in respect to movement instructive facts which help us imagine the probable process of the reversible organization and disorganization of dynamic patterns that are supposed to take place on the subcellular level.

III. Rate and Intensity

The activity of protoplasmic streaming can be quantitatively expressed in two different ways; namely, in terms of the rate or amount of protoplasm carried by the streaming, and in terms of the motive force responsible for the streaming. The former has to do with the visible end effect of the phenomenon and the latter with its immediate cause. First, we shall deal with the former.

1. Intracellular velocity distribution

In order to probe into the physical aspect of protoplasmic streaming it is extremely important to know how the rate of streaming is distributed in different layers of flow in a cell, for without knowing and analysing it, a sound approach to the physical nature of the streaming is hardly possible.

As early as 1903 EWART paid attention to this point and presented a figure of velocity distribution of protoplasm and cell sap in the internodal cell of *Chara* which shows that the streaming is greatest in the inner layers of the endoplasm and least at the interface between endoplasm and ectoplasm; some degree of streaming is also shown by the cell sap in the layers adjacent to the vacuolar membrane. It was mainly these facts which led EWART to believe that the force causing streaming is developed in the endoplasm. It is, however, unfortunate that his figure does not represent the situation of the intracellular motion correctly. After EWART's report, there were for a long time no experiments conducted along this line until KAMIYA (1950 a, 1957), KAMIYA and KURODA (1956 b, 1958 a) took up the problem recently. The materials in which the velocity distribution of the protoplasmic streaming has so far been studied in detail are *Nitella* cells and myxomycete plasmodia.

a) Rhizoid cell of *Nitella*

The rhizoid cell is especially favourable for this kind of work due to its comparatively thick plasmasol layer (endoplasm) and to the absence of chloroplasts, a fact which greatly facilitates observation. The velocities of endoplasm at different layers are measurable by numerous granules with which the endoplasm is replete and which serve as index markers. It has been pointed out by LINDENBEIN (1937), that there is no difference in rate between larger and smaller particles in the internodal cell of *Chara*. The same is true also for the *Nitella* cell. By simple observation it is recognized that all the endoplasmic particles move with a more or less equal rate along a very thin, almost imperceptible layer of cortical gel. For determining the velocities of particles in different layers of the stream at one and the same moment exactly, KAMIYA and KURODA (1956 b) resorted to the cinematographic technique.

Fig. 5 represents the state of flow in the horizontal, optical section through the axis of the rhizoid cell, the two indifferent lines being in such a position that one is at the top and the other at the bottom of the cell.

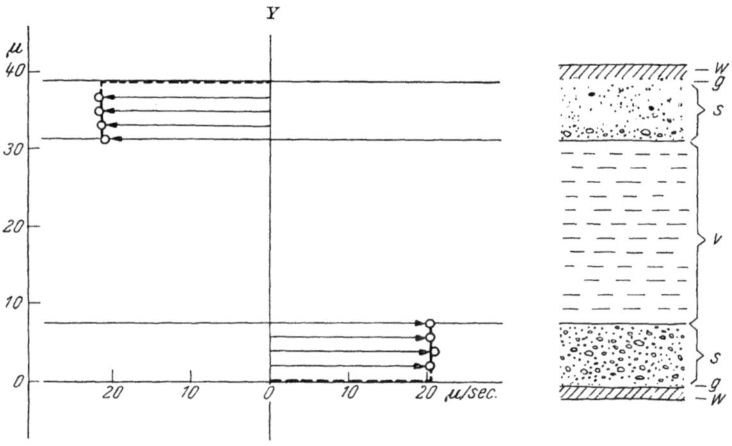

Fig. 5. Velocity distribution of the endoplasmic streaming in a rhizoid cell of *Nitella flexilis*. A part of the rhizoid cell as seen under the microscope is shown on the right. *w*: cell wall, *g*: plasmagel (ectoplasm), *s*: plasmasol (endoplasm), *v*: vacuole.
(KAMIYA and KURODA 1956 b.)

Endoplasmic granules found at a moment in an arbitrary transverse section YY' were carried along with the stream to the positions shown by the open circles, the mean rate of the granules being 21 μ/sec. This figure shows that there is hardly any appreciable velocity gradient in the moving endoplasmic layer, the whole endoplasm flowing with a nearly equal rate. The small fluctuation in rate among particles is due to their independent Brownian motion occurring in the moving endoplasmic layer. Only in the very thin layers close to the inner surface of the cortical gel an enormous velocity gradient is found. In other words, the plasmasol layer (endoplasm) is actually not streaming, but just sliding as a whole on the inner surface of the cortical layer.

b) Leaf cell of *Nitella*

The situation is the same also in "leaf" cells and in internodal cells. As a young leaf cell is rich in protoplasm taking part in the streaming and also in inclusions suspended in the cell sap, this permits us to determine the velocity distribution of both protoplasm and cell sap. Furthermore, the situation is simpler than in a grown-up internodal cell in that the course of flow is not spirally wound. From Fig. 6, which shows the motion of both protoplasm and cell sap, we realize that the rate of flow of the endoplasm is again the same irrespective of the stratum, just as in the case of the rhizoid cell in Fig. 5.

The velocity of the sap, which wets the tonoplast and is carried passively by it, decreases inward until the central axis is reached, where the velocity is nil. On the other hand, the velocity gradient of the sap, which is indicated by the slope of the curve against the transverse section

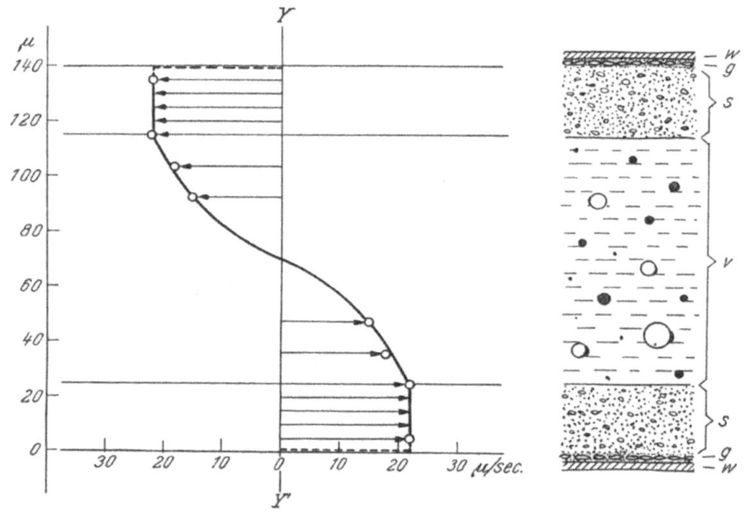

Fig. 6. Velocity distribution of the streaming of protoplasm and cell sap in a young 'leaf' cell. A part of the leaf cell as seen under the microscope is shown on the right. *w*: cell wall, *g*: plasmagel layer, *s*: plasmasol (endoplasm), *v*: vacuole.
(KAMIYA und KURODA 1956 b.)

of the cell, increases inward until the maximum is reached at the point of inflection found at the middle point of the vacuole. The shear gradients of the cell sap in the peripheral and axial regions of the vacuole are 0.25 sec.$^{-1}$ and 1.0 sec.$^{-1}$ respectively.

The slope of the curve (velocity gradient) changes abruptly at the boundary between endoplasm and cell sap, or tonoplast. This change is assumed to be caused by the difference in viscosity between endoplasm and cell sap. It is, however, not practicable in this case to evaluate the viscosity ratio between endoplasm and cell sap by comparing the velocity gradients at their boundary as there is no measurable gradient of velocity in the endoplasmic layer [3].

All in all, the form of the velocity distribution in Figs. 5 and 6 suggests that it is the sol-gel interface that plays an active part in the streaming

[3] This is presumably due to the fact that the endoplasm is possessed of a certain amount of yield value, although this may not be high.

and that the protoplasmic sol alone is rather inert in respect to movement. We shall discuss this problem more fully in Chapter IX, 1 (p. 38).

c) Plasmodial strand

The character of the velocity distribution of the protoplasmic streaming in the plasmodial strand of the myxomycete, *Physarum polycephalum,* is quite different from that in *Nitella* cells. The strand has the structure of the tube whose wall is made of gel protoplasm. By simple observation it is noticed that the speed of flow is greatest at the central region of the capillary and becomes less and less as the inner surface of the gel wall is

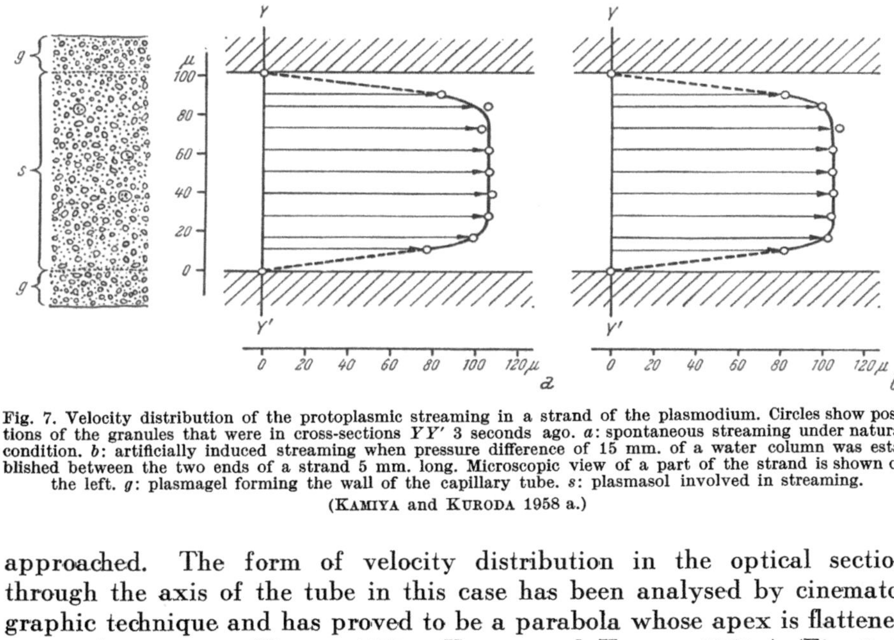

Fig. 7. Velocity distribution of the protoplasmic streaming in a strand of the plasmodium. Circles show positions of the granules that were in cross-sections *YY'* 3 seconds ago. *a*: spontaneous streaming under natural condition. *b*: artificially induced streaming when pressure difference of 15 mm. of a water column was established between the two ends of a strand 5 mm. long. Microscopic view of a part of the strand is shown on the left. *g*: plasmagel forming the wall of the capillary tube. *s*: plasmasol involved in streaming. (Kamiya and Kuroda 1958 a.)

approached. The form of velocity distribution in the optical section through the axis of the tube in this case has been analysed by cinematographic technique and has proved to be a parabola whose apex is flattened or nearly truncate (Kamiya 1950 a, Kamiya and Kuroda 1958 a) (Fig. 7 a). This characteristic form is also shared by the figure of extrusion of a fluid having structural viscosity when it is made to move through a capillary tube. Kamiya therefore came to the conclusion that the protoplasmic streaming in a strand of myxomycete plasmodium is caused passively by a pressure difference between the anterior and posterior parts of the strand. Recently Kamiya and Kuroda (1958 a) confirmed this conclusion by further experiments to which we shall refer in a later chapter (p. 142).

Locquin (1949) stressed that the moving protoplasmic portion in a plasmodial capillary of *Fuligo* keeps a perfect cohesion and is displaced in bloc. This is true as far as the axial region represented by the flat part of the velocity distribution is concerned. This flat part gets narrower as the rate of flow increases. Hence, there is no general validity in Locquin's statement.

2. Rate

It is the rate that has been almost exclusively used by many authors as a criterion of the activity of the protoplasmic streaming. For the speed of flow is readily measured and there have been no other practical quantities representing the strength of the flow.

The rate of protoplasmic streaming varies over a wide range from an almost imperceptible speed to more than 1000 μ/sec. according to the material and also to the external conditions. The rate may vary also from cell to cell in the same tissue. Even in a given cell the speed of flow may not necessarily be the same at different parts of the stream. Furthermore, in many plant cells, *e. g., Elodea* or *Vallisneria,* it often happens that a larger granule like a chloroplast moves more slowly than a smaller granule.

a) Methods of measurement

Measurement of the absolute rate of protoplasmic streaming can be made simply by timing the movement of particles between two selected points in the standardized eye-piece scale by means of a stopwatch or by an appropriate sound signal. STRUGGER (1949) states the statistical treatment of the data thus obtained. This method suffices for ordinary objects or purposes, but other ways of measurement were also developed by several workers in order to obtain more objective and accurate values of the rate. They are classified into two types; namely, one based on comparison with a stream of a variable but known velocity (comparison method), and the other involving direct optical recording of the movement.

Comparison methods. If an image of a strip of a running belt set up outside the microscope is observed through a camera lucida together with the protoplasmic streaming and if the speed of the belt is controlled either manually or electrically so that the rate of protoplasmic streaming and the movement of the belt become the same, we can determine the rate of streaming from the optical magnification and the speed of the running belt. This type of method was adopted by SWEENEY and THIMANN (1942) for coleoptile cells of *Avena,* by KAMIYA (1950 a) for plasmodia of *Physarum,* by KELSO and TURNER (1955) for staminal hair cells of *Tradescantia.* It is especially advantageous to use such a method for determining the time course of the ever-changing rate of streaming in myxomycete plasmodia.

Photographic recording methods. There are some techniques developed to register the motion of protoplasmic granules directly on a photographic film. One of the merits of this technique is its ruling out personal error in observation. Under suitable dark field illumination, granules carried along with the flowing protoplasm are recognized as bright spots. If under this condition, the picture is taken with an exposure of a definite duration, these bright spots leave streaks on the film corresponding to the distances they traverse during that interval. The exposure time and length of a streak being known or measurable, one is ready to calculate the absolute speed of the granule from the record. This is a well-known technique in hydraulics or aerodynamics which was applied by DUBUY and OLSON (1940) to the determination of the rate of protoplasmic flow in coleoptile cells of *Avena*. Application of this method to the study of protoplasmic streaming, however, encounters some difficulties in that well-defined streaks thus obtained usually represent distances covered by comparatively larger granules or lipid drops which do not always show faithfully the rate of the streaming of the ground plasm (cf. SWEENEY and THIMANN 1942).

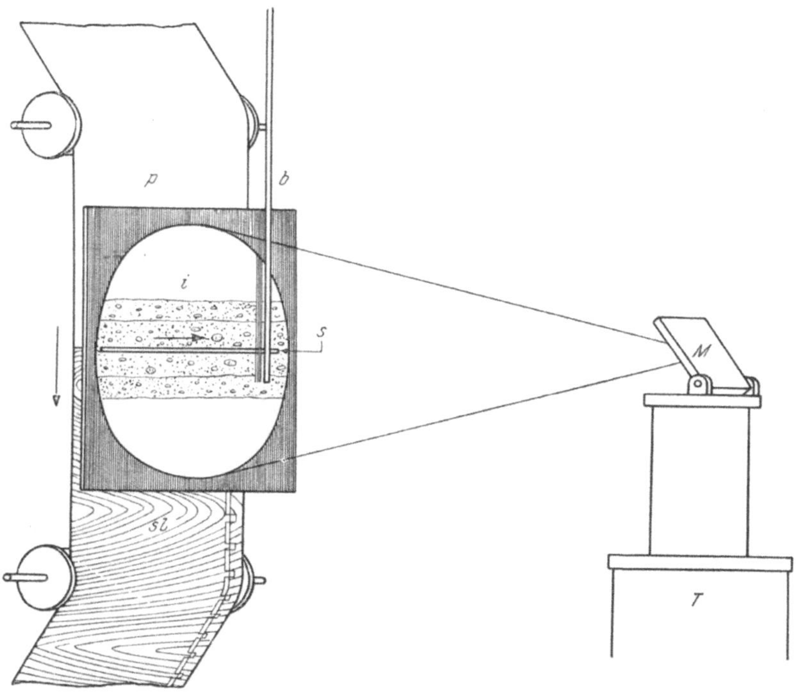

Fig. 8. Set-up for recording shadow lines of the protoplasmic streaming. *T*: microscope tube, *M*: surface mirror, *i*: image of the plasmodial strand projected on the screen of the photokymograph, *s*: horizontal slit, *p*: photographic film or paper running downward, *sl*: latent image of the shadow lines, *b*: bar serving as a time marker.
(Original.)

There is another method of photographic registration which was called "shadow imprinting method" by KAMIYA (1950 a) and was applied to the study of successive changes in the rate of the spontaneous rhythmic streaming in a plasmodial strand of *Physarum polycephalum*. The principle of the method is the same as that of the method used by HÜRTHLE (1915) for determining the rate of blood flow in mesenterium capillaries of the frog. If the axial part of a straight strand of the plasmodium is projected on a slit of a photokymograph, the light passing through the slit will fall on the film behind the slit as a thin line consisting of spots of different light and shade owing to the optical heterogeneity of the endoplasm (Fig. 8). As the endoplasm moves, so does the image of light and shade spots on the film accordingly. If the film is made to move with a constant speed at a right angle to the slit, light and dark lines will be traced running more or less obliquely according to the rate of flow and

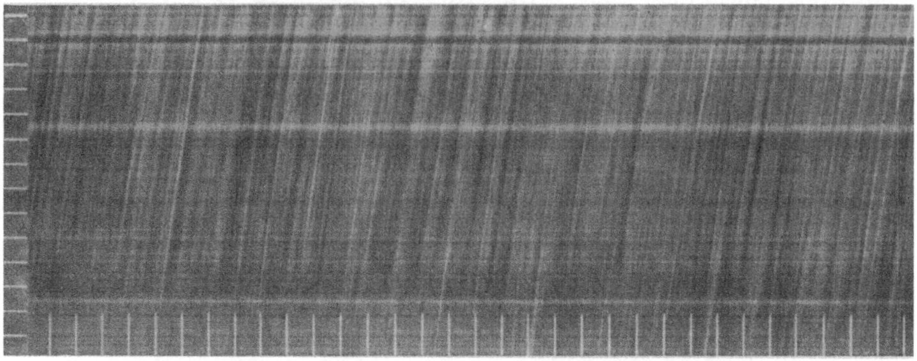

Fig. 9. Shadow lines of rotational streaming in an internodal cell of *Nitella flexilis*. A unit interval of the graduation at the right end: 20 μ, time interval: 6 seconds.

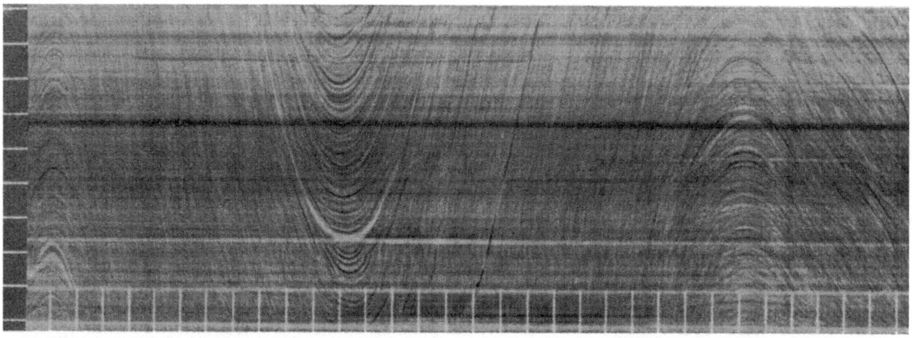

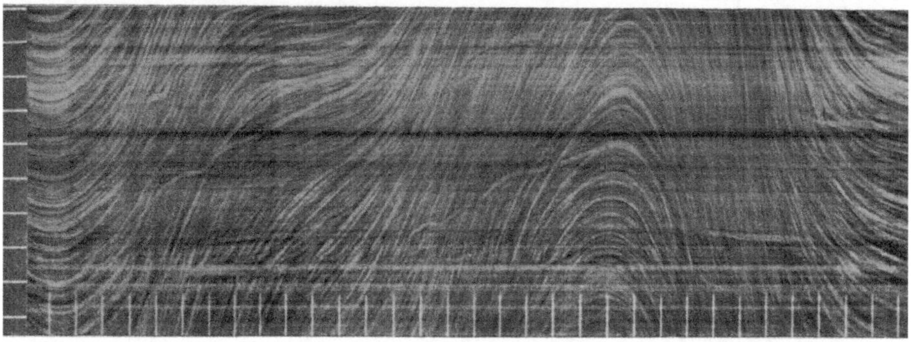

Figs. 10–11. Shadow lines of the shuttle streaming in plasmodial strands of *Physarum polycephalum*. Figs. 10 (top) and 11 (bottom) were obtained from a comparatively thick and thin straight strands respectively. A unit interval of the graduations at the right end: 50 μ, time interval: 6 seconds.

(Original.)

the speed of the film. This inclination represents the rate at which the endoplasmic granules flow.

Fig. 9 represents a record obtained from a rhizoid of *Nitella flexilis*.

We see that imprint shadow lines run here straight and almost all of them go in parallel as though rubbed with a brush. This is a good proof that all particles whose shadows reached the photographic paper through the slit of the kymograph move with a nearly equal and constant rate over a stretch of about 300 μ corresponding to the width of the strip of the photographic record. From this inclination we know that the absolute rate of this flow was 25 μ/sec.

Fig. 10 is an example of the shadow lines of protoplasmic streaming in a comparatively thick plasmodial strand. At each reversal of the direction of flow, the inclination of the shadow lines reverses also. The steeper the inclination is against the horizontal axis, the faster is the flow. At the point where the tangent of the shadow line becomes zero, the flow comes to a standstill. This point thus represents the time when the reversal in the streaming direction took place.

Fig. 11 is the record of the flow in a comparatively thin straight capillary. This is an example showing that the streaming rate at the center of the same capillary varies according to regions, a fact indicated by different inclinations of the shadow lines at the same time coordinate at different levels of the photographic strip. The reversal of the direction takes place in this example by no means simultaneously throughout the capillary, but is transmitted from one end to the other.

The time course of the change in the rate of flow in *Physarum* as determined by the shadow imprinting method will be shown later (Fig. 17, p. 29). Naturally, this method finds its adequate application in a variety of protoplasmic motions, but it is especially advantageous for determining the rate of flow which is exceedingly high and variable as in a plasmodium. The shadow imprinting method, furthermore, needs no strong illumination as is the case in dark field illumination applied by Olson and DuBuy (1940).

b) Size of particles and its relation to their rate of movement

It has been repeatedly noticed in not a few plant cells that smaller cytoplasmic particles move with more constant and greater rate than larger ones such as chloroplasts. Zurzycki (1951) measured in an *Elodea* cell the rates of chloroplast movement and of the streaming of cytoplasm simultaneously. The latter speed was determined by minute granules suspended in it. It was shown clearly in Fig. 12 that all chloroplasts move more slowly than the more minute particles in the cytoplasm. The figure reveals that the movement of chloroplasts is not only slower but is more irregular than that of the smaller cytoplasmic particles. This is probably due to the fact that small particles meet fewer obstacles than larger particles. The motion of chloroplasts is sometimes brought to a stop already after 2–3 hours while the protoplasm is still rotating rapidly for quite a long time. When the protoplasm takes little part in the streaming, the chloroplasts also remain in rest: the stronger the streaming becomes, the more chloroplasts take part in the motion. As Fitting (1925, p. 329) describes it, it seems as if the chloroplasts are more or less anchored firmly to the external, immobile, and more viscous plasmic layers of the cell. Once the cell is excited, the streaming spreads first to these layers so as to set the chloroplasts in motion. We can say that in many plant cells, *e. g.*, *Elodea*, *Vallisneria*, etc., smaller particles represent the rate of flow of the ground plasm more accurately than larger ones. The irregular results of observations of rotation (*e. g.*, JURIŠIĆ 1925) are likely, as pointed out by ZURZYCKI (1951), to be due to the fact that they were obtained from the rate of movement of chloroplasts.

BĚLEHRADEK (1925) referred to the ratio between the speed of chloroplasts (*Vc*) and speed of protoplasm (*Vp*), which is represented by that of smaller particles, or *Vc*/*Vp*, as an index of viscosity. He believed that chloroplasts

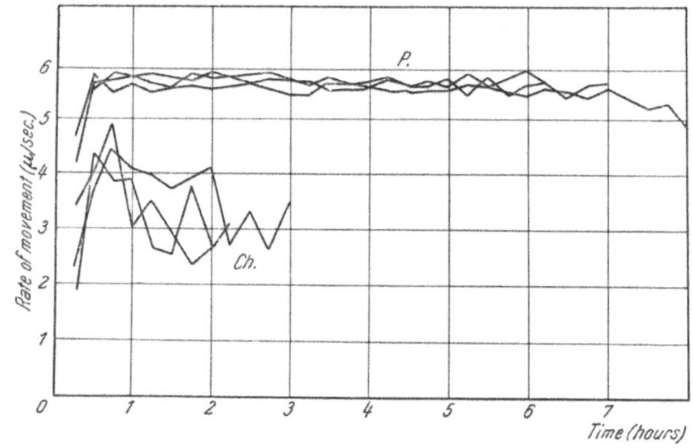

Fig. 12. Rates of movement of minute particles (*P*) and chloroplasts (*Ch*) in an *Elodea* cell under constant environmental conditions (20° C).
(ZURZYCKI 1951.)

move together with protoplasm when the viscosity of the latter is high, but the rate of the movement of chloroplasts becomes smaller when the viscosity of protoplasm is lowered. An important factor which has a direct bearing on the "index of viscosity" is, however, likely to be the friction between chloroplasts and the inner surface of the resting cortical layer. It is doubtful whether such a ratio of speeds is useful as a reliable criterion of the viscosity of the streaming protoplasm (cf. HEILBRUNN 1928, p. WEBER 1927 a).

When grown-up cells of the *Elodea* leaf are observed under the centrifuge-microscope at around 2500 g, we notice that the cell contents are stratified into three distinct layers: first, the green, opaque layer closely packed with chloroplasts and minute granules, second, the transparent layer of protoplasm (hyaloplasm) and third, the vacuole occupying around 90% of the whole cell space. As the boundary between the cell sap and hyaloplasmic layer is quite flat when the cell is being centrifuged at high acceleration, we can readily estimate from the stratified figure the volume of endoplasmic sol involved in streaming and the total volume of chloroplasts carried with the streaming. We thus realize that the volume of endoplasmic sol is about one half of the total volume of chloroplasts. This means that the streaming layer must be very thin, much thinner than the height of a chloroplast, a fact which is also directly recognizable under high magnification. Anyway, it is understandable from this fact that chloroplasts are brought into contact with the cortical gel layer, and hence face a stronger resistance to motion than the smaller particles which are freely suspended in the flowing endoplasmic layer.

In the case of *Nitella* or *Chara* cells, on the other hand, the situation is quite different. Though most of the chloroplasts are mounted in the cortical layer of the cell, there are in these cells also some which are freely suspended in the endoplasmic layer and move around with the streaming. There is, however, no distinct difference in the streaming rate between chloroplasts and smaller particles as has been pointed out by LINDENBEIN (1937). Beautiful parallelism of the shadow lines as seen in Fig. 9 also speaks for this fact, since if different particles move with different rates shadow lines are no longer in parallel but cross one another. That the amount of moving endoplasm is sufficiently large so that the chloroplasts are freely suspended in it probably renders the rates of movement of all the particles, larger and smaller, the same as the rate of flow of the matrix, *i. e.*, the hyaloplasm.

c) Length of the cell and its bearing to the rate of flow

For studying the relation between the cell length and the rate of streaming, materials, such as root hairs, pollen tubes etc., are favourable, as they have diverse lengths with a nearly constant diameter. It is known that the rate of streaming in a root hair is generally small when the hair is very short and increases gradually as it elongates until the hair reaches a certain length beyond which the rate remains constant (VESQUE-PÜTTLINGEN 1876). DOI (1950) found in his experiment, which was carefully designed to avoid mechanical injury, that the rate of flow in root hairs of *Oryza sativa* was constant at 9.1 μ/sec. at 20^0 C, independently of the length of the hairs if they were longer than 200 μ. Below 200 μ, the rate decreases. A similar relation is also found in *Triticum* root hairs. According to IWANAMI (1952, 1956) the rate of protoplasmic streaming in pollen tubes of *Lilium auratum* changes in relation to the length of the tube, but it becomes almost constant after the tube has attained a length of about 2,000 μ.

In these observations, however, it is difficult to analyse whether the dependence of the rate upon the cell length might really be due to the difference in the physical factor of cell length alone or to the difference in more intrinsic properties of the protoplasm itself which might be altered in the course of cell development. For studying the relation between the rate of flow and the cell length there is another method of approach. Protoplasmic streaming in a long cell, especially the internodal cell of *Nitella* or *Chara* can be readily tied off into two or many fragments with a strip of fine thread. The tied-off cell fragment survives quite normally, even when the adjacent cell fragment on the other side of the ligature is cut off (cf. DUTROCHET 1837). LINSBAUER (1929), who studied in detail the effect of ligation of the cell, ascertained that new cellulose septa are formed on both sides of the tied-off region. There is no doubt left especially in the light of SANDAN's experiment (1955) demonstrating that a cell fragment, both ends of which are tied off, can develop new shoots and rhizoids and behaves as a completely independent cell. At any rate, the protoplasmic streaming in the cell fragments thus made stops at first for a brief period after the operation due to the shock, but it is resumed in about 10 minutes (cf. KURODA 1956) (Fig. 13). What is then the rate of resumed flow in relation to the length of the cell fragment? HAYASHI (1952) found that the rate of endoplasmic flow is the same as that of the normal cell before the operation was conducted, provided the length of the cell is not less than 1 mm. Below 1 mm. the rate diminishes rapidly (Fig. 14). What this fact implies is that the active places generating the protoplasmic flow are not confined to some special loci of the cell in relation to its longitudinal axis, *e. g.*, to the terminal ends of the cell, but is distributed evenly along the entire length of the cell.

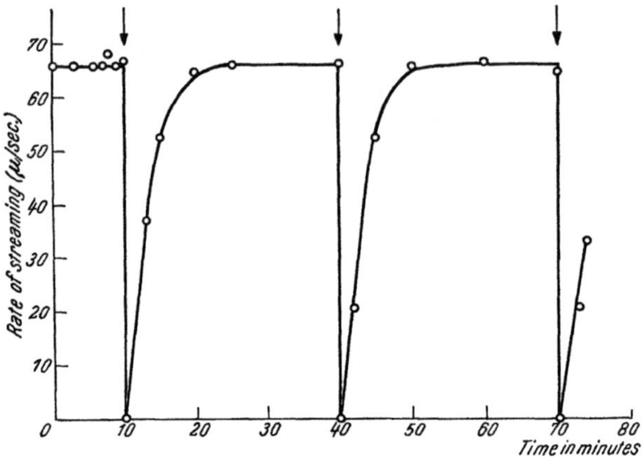

Fig. 13. Restoration of the rate of protoplasmic streaming after the operation of ligation followed by amputation. Abscissas: time in minutes, ordinates: rate of flow. Operations were conducted at the time indicated by arrows. The length of the cell was diminished from 31 mm. to 16 mm. by the first operation, to 13 mm by the second, and to 6 mm. by the third operation.
(KURODA 1956.)

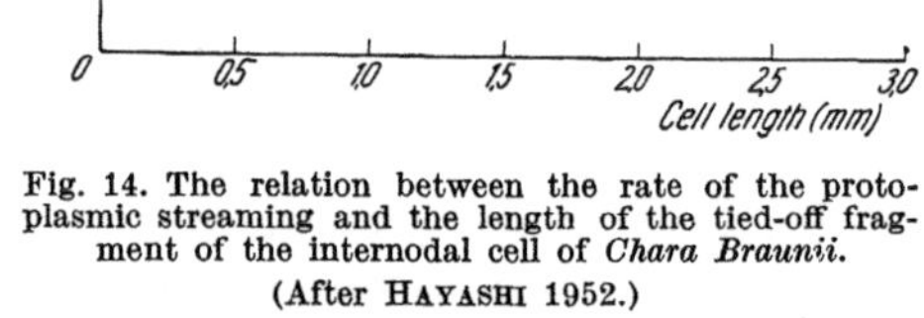

Fig. 14. The relation between the rate of the protoplasmic streaming and the length of the tied-off fragment of the internodal cell of *Chara Braunii*.
(After HAYASHI 1952.)

d) Spontaneous changes in the rate of flow in myxomycete plasmodium

The protoplasmic streaming in myxomycete plasmodia is unique and of special interest in that its rate is not only enormously high but is also ever accelerated or decelerated rhythmically under a constant external condition according to an autonomous scheme. One should be informed as a piece of basic knowledge as to how this speed keeps changing with time. We have already dealt with how the rate of flow is distributed in a plasmodial strand at an arbitrary moment. Here we shall deal with the time course of the spontaneous change in the rate of flow at the axial region of the intracapillary stream. In order to follow it, instantaneous values of speed must be determined without interruption. For this purpose KAMIYA (1950 b) made use of the two techniques described before, *i. e.*, the comparison method and the shadow imprinting method.

For a graphical representation of the streaming velocity, either one of the two directions is taken as (+) and the other as (—). When a series of velocity values with (+) and (—) signs are plotted on graphs as ordinates against time as abscissas, undulating curves are obtained which denote autonomous changes in the rate and direction of the protoplasmic streaming.

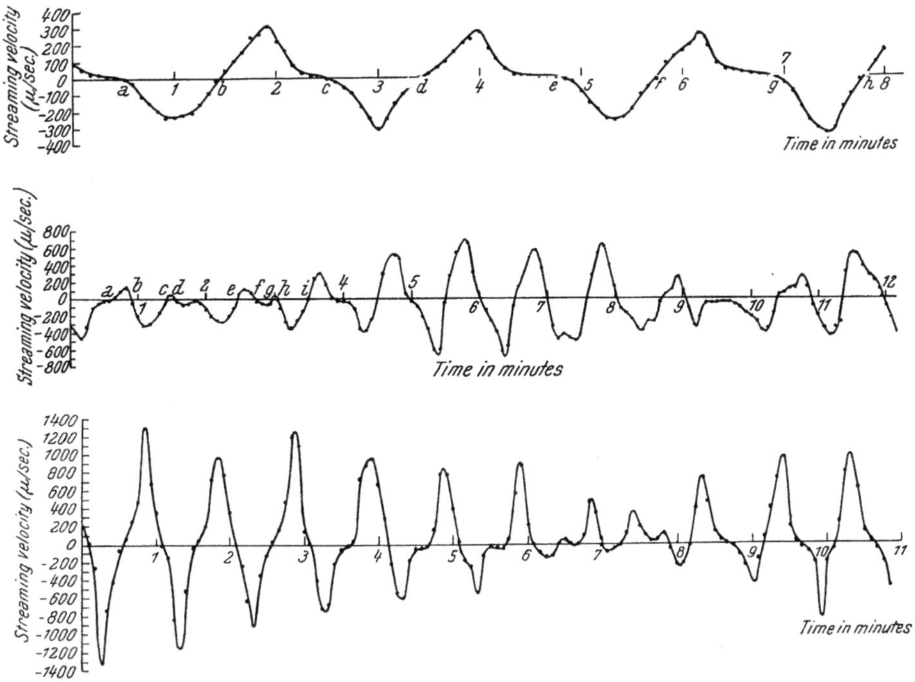

Figs. 15–17. Spontaneous changes in the rate of protoplasmic streaming in plasmodial strands *(Physarum polycephalum)*.
(KAMIYA 1950 b.)

The wave train shown in Fig. 15 is obtained from an experiment conducted on *Physarum polycephalum* at 16.5° C. The point at which the curve intersects the base-line indicates the moment of reversal of the streaming direction. VOUK (1910) referred to the time it takes for the protoplasm to flow in one direction and next in the other direction as rhythm duration (*Rhythmusdauer*). This is represented here by the time interval *a–c* (121 sec.), *b–d* (119 sec.), *c–e* (129 sec.), etc. The values which can be measured easily by means of a stopwatch are in general only those of the streaming duration (*Strömungsdauer*, VOUK 1910) such as *a–b* (57 sec.), *b–c* (64 sec.), *c–d* (55 sec.), etc. or the rhythm duration just mentioned. As is clear from this figure, the streaming, after reversing its direction at the time *a*, increases its speed in the minus direction till it reaches 220 μ/sec., then slows down again to stop finally at *b* (1 min. 26 sec.). There it starts immediately going in the plus direction, its velocity increasing almost linearly till it attains the maximum (300 μ/sec.) after which the curve takes again a downward course. In so doing the velocity decreases first precipitously, then gently and finally becomes zero at *c*, after which the flow starts going again in the negative direction. Such a back and forth movement is repeated by the protoplasm of the slime mould as long as it is active. In this experiment the average rhythm duration was 2 minutes 7 seconds and the maximum rate of flow during 8 minutes was 320 μ/sec. All of the three waves seen in the figure present characteristic asymmetrical forms which are very similar to one another.

The pattern of the velocity curves, however, keeps changing from wave to wave even at the same locus of the same plasmodium. In Fig. 16 (temp. 29° C.) a characteristic wave form keeps changing little by little at each wave. The amplitude, small at first, becomes reinforced 5–8 minutes before it goes diminishing. The maximum speed reaches ca. 700 μ/sec. in both plus and minus directions. Special attention should be called here to a couple of two-peaked waves at the beginning. The reversal of the flowing direction takes place at the time coordinate *a, b, c, ...* designated in the figure. At *f, g,* and *h* the protoplasm reverses its streaming direction after an extraordinarily short lapse of time, for *e–f, f–g, g–h, h–i* amount to 17, 13, 5, and 28 seconds, respectively. If one were to base one's judgement of the rhythm of protoplasm merely upon the reversal of the flowing direction, one would necessarily be led to believe that the protoplasmic rhythm is sometimes very much shortened, when actually we are dealing with two-peaked waves. The nature of rhythmicity will be discussed in detail in chapter V in connection with motive force curves.

Fig. 17 (temp. 28.4° C.), which was obtained by means of the shadow imprinting method, shows a record from a very active material. The average rhythm duration is 60 seconds, that is, less than the half of that in Fig. 15. The waves have a slightly gentler inclination near the base-line, but present an exceedingly acute form in both plus and minus sides midway between the two crossing points.

Besides the rhythmic reversal, the enormous speed as compared to the speed of rotation in plant cells is also one of the significant features of the protoplasmic flow in the slime mould. Vouk obtained 1250 μ/sec. as the maximum value of the streaming rate in *Didymium nigripes.* In Fig. 17 the maximum and minimum values vary greatly in each wave, reaching 1350 μ/sec. (at 50 sec.) and 1300 μ/sec. (at 2 min. 52 sec.). These values surpass the highest rate ever known for the velocity of protoplasmic flow. In *Physarum polycephalum,* it is not rare that the streaming velocity surpasses 1000 μ/sec. at optimum temperature. In determining high rates of flow of more than 1000 μ/sec., however, we cannot expect to obtain sufficiently reliable values from the routine method using stopwatch and ocular micrometer, for it becomes increasingly difficult to follow the streaming granules under the microscope. It is a great advantage of the shadow imprinting method that it enables one to determine the speed of rapidly flowing protoplasm which is too fast for microscopic observation. Though the speed of 1350 μ/sec. is the highest so far encountered, it is most likely that in vigorous strands, which are usually too opaque to permit shadow registration, the protoplasmic streaming takes place with still higher speed. KAMIYA (1942) noticed a case where the speed of flow was artificially accelerated beyond 2000 μ/sec. without inducing any pathological effects when a pressure difference was established between two different portions of one and the same plasmodium by means of the double-chamber method (see p. 40).

On inspecting the graphs representing the spontaneous rhythmic change in the streaming rate, we realize that the patterns of the curves are all different according to the material used. At the same locus of the capillary, an identical rhythmic change in speed is usually not repeated, although close relationship is found between successive waves both in amplitude and wave form. The maximum of the curve is not always found in the middle of the streaming duration, *i. e.*, between two adjacent crossing points. Moreover, one streaming duration has occasionally more than one maximum and minimum of velocity.

It has been an implicit custom among investigators to use the moment of reversal of flow as the criterion for separating one rhythm from another. But the moments of reversal, where the motive force is zero, do not necessarily imply the same phase of rhythm. The statement of VOUK (1910) that the rhythm duration, the duration covering the successive forward and backward streamings, is nearly constant holds true only when the velocity wave is regular. When a wave shows an irregular and somewhat complicated pattern which appears usually in periods having a diminished amplitude, this view cannot be held any longer. The two-peaked waves in Fig. 16 are a striking case in this respect.

e) Rate of protoplasmic streaming in various materials

The rate of protoplasmic flow is very different in various materials. It also changes to a great extent according to the internal conditions of a cell even when external conditions are kept the same. Therefore the comparison of the rates among different materials as listed in Table 1 is not to be taken too strictly, but it will be a help in getting an idea as to the rate of flow among diverse materials. This table is elaborated on the basis of the tables presented by BUSHEE (1908) and by VOUK (1913) by adding or substituting data from later authors.

3. Course and direction of rotational streaming

In cells of the *Elodea* or *Vallisneria* leaf, root hairs, pollen tubes and in many other long cells, streaming takes place usually along the longitudinal axis of the cell. It was believed that the course of streaming is determined so as to encounter "least resistance" (cf. VELTEN 1873, p. 86; BERTHOLD 1886, p. 121; HÖRMANN 1898, p. 16; PFEFFER 1904, p. 729; KÜSTER 1956, p. 66). This is, however, not an absolute rule. In adult internodal cells of *Chara* or *Nitella* protoplasm flows along a spiral course. Perhaps there are close

Table 1. *Rate of protoplasmic streaming in various plant materials*

Plant	Organ	Rate (μ/sec.)	Author	Temperature (°C.)
Didymium nigripes	plasmodium	1250	VOUK (1913)	
Didymium serpula	plasmodium	170	HOFMEISTER (1867)	
Physarum polycephalum	plasmodium	1350	KAMIYA (1950 b)	28
Rhizopus nigricans	hypha	55	ARTHUR (1897)	28
Mucor stolonifer	hypha	55	SCHRÖTER (1905)	26–28
Phycomyces nitens	hypha	55	SCHRÖTER (1905)	26–28
Phycomyces Blakesleeanus	sporangiophore	2–4	POP (1938)	
Humaria leucoloma	hypha	250	JAHN (1934)	
Caulerpa prolifera	'leaf'	2–3	DOSTAL (1929)	12–30
Acetabularia calyculus	Stalk	2–5	original	25
Nitella mucronata	'whorled leaf'	54	LAMBERS (1925)	20
Nitella mucronata	internode	52	UMRATH (1934)	20
Nitella flexilis	internode	78	KAMIYA and KURODA (1956)	28
Nitella flexilis	rhizoid	21	KAMIYA and KURODA (1956)	23
Chara foetida	internode	42	VELTEN (1876 a)	20
Chara Braunii	internode	75	HAYASHI (1952)	27
Elodea canadensis	leaf	10	BUSHEE (1908)	20
Elodea densa	leaf	6	ZURZYCKI (1951)	20
Vallisneria spiralis	leaf	10–15	JURIŠIĆ (1925)	18
Tradescantia virginica	staminal hair	5.4	KELSO and TURNER (1955)	24
Cucurbita maxima	petiole hair	4,3	BUSHEE (1908)	20
Allium cepa	inner epidermis	4	JAROSCH (1956 b)	
Avena sativa	coleoptile epidermis	10	BOTTELIER (1934)	21
Avena sativa	root hair	5.4	BUSHEE (1908)	20
Gloxinia speciosa	leaf blade hair	5.3	BUSHEE (1908)	20
Hydrocharis morsus ranae	root hair	8.1	HOFMEISTER (1867)	
Trianea bogotensis	root hair	9.5	JURISIC (1925)	18
Triticum aestivum	root hair	7.4	DOI (1950)	20
Oryza sativa	root hair	6.7	DOI (1950)	20
Hordeum vulgare	root hair	7.0	DOI (1950)	20
Zea Mays	root hair	5.7	DOI (1950)	20
Raphanus sativus	root hair	4.8	DOI (1950)	20
Cucumis sativus	root hair	4.8	DOI (1950)	20
Lycopersicum esculentum	root hair	5.9	DOI (1950)	20
Spinacia oleracea	root hair	4.9	DOI (1950)	20
Lilium longiflorum	pollen tube	5.5	IWANAMI (1953)	28
Lilium auratum	pollen tube	4.3	IWANAMI (1953)	26
Oenothera lamarckiana	pollen tube	5.5	IWANAMI (1953)	26
Vicia Faba	pollen tube	2.9	IWANAMI (1953)	20
Pisum sativum	pollen tube	2.3	IWANAMI (1953)	24

connections among the direction of the streaming, the structure of the cortical gel and the stage of cytogenesis.

The orientation of streaming courses or tracks is often modified by artificial means, such as a wound [NICHOLS 1925 (*Nitella*); KIRCHHEIMER 1933 (*Phycomyses*); JANSE 1890, DOSTAL 1929 (*Caulerpa*)], plasmolysis [KÜSTER 1910 (*Elodea*), HAYASHI and KAMITSUBO (*Nitella*) (p. 98)] bending or folding [KÜSTER 1956 (*Codium*), IWANAMI 1952, 1956 (pollen tube, *Lilium*)] etc. Fig. 18 shows how the protoplasmic streaming in the tortuous pollen tube of *Lilium* takes the shortest path when it is mechanically folded. We are inclined to believe from such a figure that the thread consisting of streaming endoplasm and cortical plasm is stretched and therefore is to be considered more or less elastic.

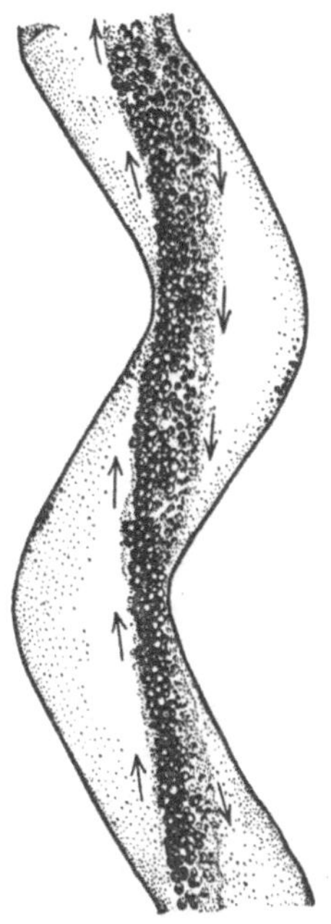

Fig. 18. The path of the protoplasmic streaming in a tortuous pollen tube (*Lilium auratum*). (IWANAMI 1956.)

The nature of the factors lying in the protoplasm that decide whether it will go clockwise or counter-clockwise in a definite space is another problem. In the streaming in *Characeae*, the representative material of the 'primary streaming,' the direction of flow is regularly interrelated with the morphological position of the cell as studied by BRAUN as early as 1852 (cf. KÜSTER 1956, p. 68). The streaming direction in *Characeae* is very stable and is never reversed throughout the whole life of the cell under normal conditions. In contrast to this, the direction of the 'secondary streaming,' as exemplified by *Elodea* leaf, is unsettled. As mentioned before (p. 14), rotation is established *via* agitation and circulation. As to what the direction is that the rotation will eventually take, whether it will go clockwise or counter-clockwise, we assume that chance will decide it. BERTHOLD (1886) pointed out that in *Elodea* cells the protoplasmic streaming, which has been brought to a standstill through mechanical means, recurs after a while in such a way that sometimes two streamings occur in one and the same cell in opposite directions until one is eventually overcome by the other and joins it. EWART (1903) states that change in direction of streaming can sometimes be observed in an *Elodea* leaf if stimuli are applied which are sufficiently great to produce death in some cells and temporary stoppage of streaming in others. KUSUNOKI (1940) observed that the streaming direction of protoplasm is occasionally reversed in *Hydrilla* cells after the leaf is dried to a considerable extent. In an experiment of ultracentrifugation of *Elodea* cells (BEAMS 1949), about 7% of the total cells reversed the direction of flow after the entire cell contents, except vacuoles, were strongly pushed down in the centrifugal field (cf. p. 103). There is as yet no experimental procedure known to produce consistent reversal in the streaming of *Elodea* cells.

It has been pointed out that the direction of streaming is determined not only by the factors within the cell but also by the mutual influence of the adjacent cells. That such a relation exists between neighbouring cells has been shown statistically by PEKAREK and FÜRTH (1931). More often than not the directions of the rotational streamings in the cells adjacent to a definite cell are found to be the same as that of the latter cell. When the streaming in *Elodea* cells is brought to a standstill by one way or another, and later starts again, then it takes the other direction in many cells. Such cells are, however, always situated at the boundary region between the groups of cells in which the streaming direction remained unchanged. It does not happen that a cell in the midst of a cell group, where the streaming direction is the same in every cell, changes its streaming direction independently of the others under new inducement.

4. Transport-volume and intensity of flow

As a rule we distinguish intensity of flow from its rate. The significance of protoplasmic streaming has been pointed out by several workers (BOTTELIER 1934, VELTEN 1872) from the viewpoint that the protoplasmic streaming would play a significant part for intracellular transportation of substances or for the locomotion of the organism (slime mould). Since no explicit definition of the intensity of the streaming had been given, KAMIYA (1950 c) defined it as the volume of protoplasm passing an arbitrary cross section of a stream channel or of a cell per unit time. Even when protoplasm moves in different directions simultaneously through the section under consideration, the total arithmetical sum of the transport-volume per unit time is the measure of the intensity of flow.

Although it is not at all simple to obtain the absolute amount of intensity of the protoplasmic streaming in general, we can calculate or evaluate it in special cases. We shall consider this amount in the following. In order to evaluate it directly, we must know the volume of protoplasm involved in the streaming and its mean velocity. For this purpose we must first determine the velocity distribution of the flow.

a) Flow intensity in *Nitella* cells

Fortunately, there are, as already referred to, some data concerning the velocity distribution of the rotationary streaming in *Nitella* cells (KAMIYA and KURODA 1956 b), data which are available for evaluating the transport volume of protoplasm of the rotational streaming. We know in this case rather exactly the diameter of the cell, the thickness of endoplasm involved in the streaming and the rates in its different layers, which, however, are regarded as being equal (Figs. 5 and 6). Thus we get the intensity of the flow in a rhizoid cell of *Nitella flexilis;* this is 37,000 μ^3/sec. ($= 0.37 \times 10^{-4}$ μl./sec.). That in a leaf cell of the same plant is 205,000 μ^3/sec. ($= 2.05 \times 10^{-4}$ μl./sec.). As an equal amount of protoplasm is carried in opposite directions, the amount of protoplasm flowing in one direction is half as much as that in each of the above cases.

b) Intensity of intracapillary streaming in the myxomycete plasmodium

We can evaluate the transport-volume per unit time of the streaming along a strand of a slime mould if the intracapillary velocity distribution is determined at a time. However, inasmuch as the streaming in most slime moulds is always accelerated or decelerated and is never in a stationary state, the velocity distribution must be determined successively if we try to figure out how the intensity of flow changes in relation to time. This is very tedious, if not impossible. KAMIYA (1950 c) obtained the spontaneous change in intensity of the protoplasmic flow in a plasmodial strand by means of an entirely different method.

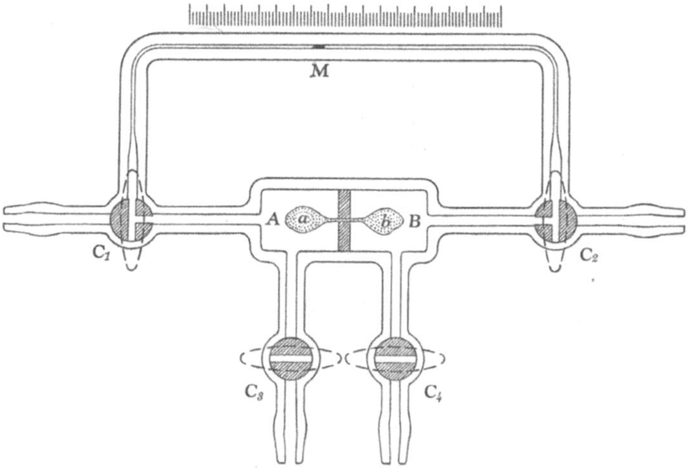

Fig 19. Double-chamber volumeter. *A*, *B*: compartments containing small masses of protoplasm *a* und *b*. *M*: index drop. C_1–C_4: stop-cocks. (KAMIYA 1950 c.)

(1) *Transport volume.* We shall first see how the protoplasmic streaming in a myxomycete plasmodium behaves when measured by means of a volumetric method.

Since in the myxomycete plasmodium (*Physarum polycephalum*), as stated in the foregoing chapter (p. 13), the protoplasm streams back and forth alternately, it is the net difference between the quantities of protoplasm carried in two opposite directions which contributes to its locomotion. It is clear that, in so doing, the plasmodium advances in that direction in which a larger amount of protoplam is transported. The flow in the direction in which the plasmodium advances is called 'progressive streaming' and that in the opposite direction 'regressive streaming' (VOUK 1910). Hence the former carries obviously a greater quantity of protoplasm than the latter.

KAMIYA attempted to measure the endoplasmic flow and the consequent locomotion of the organism in terms of the volume of protoplasm transported by the streaming, or 'transport-volume.' One of the volumeters devised by him for this purpose is shown in Fig. 19. The construction of the chamber is such that the central partition separates the chamber into two air-tight compartments while the protoplasmic flow takes place freely between the two blobs along the connecting strand which penetrates the central partition.

It is obvious that when the endoplasmic flow takes place from *a* to *b*, the volume of *a* decreases to the same extent as the volume of *b* increases; when the protoplasm streams in the opposite direction, the situation will be reversed. This change in volume induced by the streaming will be transmitted by the air to the index-marker in the glass capillary. In the experiment, the successive position of one of the menisci of the index-drop is read by means of a measuring microscope at definite intervals or registered automatically on a strip of photographic paper which is moving at a right angle to the capillary at a constant speed.

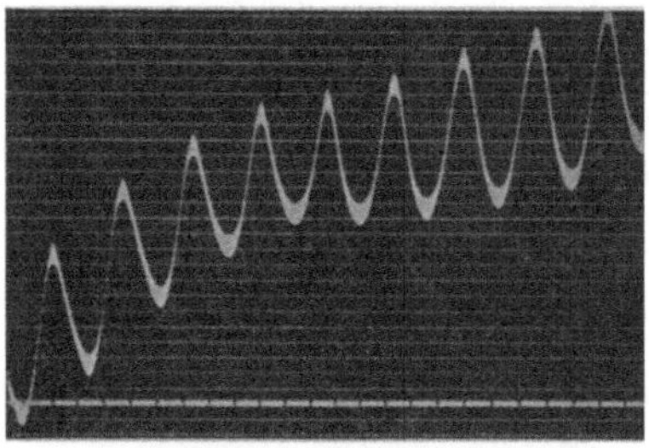

F g. 20. Volumetric registration of the protoplasmic streaming in the myxomycete plasmodium. Unit interval of the parallel lines: 0.2 µl., time interval: 30 seconds. (Kamiya 1950 c.)

The fact that protoplasm flows somewhat more in the direction in which the whole body advances than in the opposite direction can be clearly shown on a graph thus obtained.

Fig. 20 represents an example of automatic registrations of the protoplasmic streaming in terms of the transport-volume. The interval between two parallel lines has been set up in the experiment in such a degree of magnification so that it is just 0.2 μl.; the time interval is 30 seconds. The time coordinate of a crest of the wave shows the instant at which the flow from *a* to *b* is over and the direction is being reversed to from *b* to *a*: that of a trough shows the moment at which the flow from *b* to *a* is being reversed to from *a* to *b*.

The waves obtained by the above method represent the cumulative effect of transportation by the flow. While the cumulative effect stands in itself for the locomotion of the plasmodium, the rate of transportation will be the measure of the *intensity* of the flow.

(2) *Intensity of the intracapillary streaming.* As has been shown in Fig. 7 *a* (p. 21), the intracapillary velocity distribution of the protoplasm approaches the revolution of a parabola, the apex of which is more or less flattened. Referring to Fig. 7, particles in the protoplasm that are placed within a section plane YY' at a given time t move at the next moment $t + \Delta t$ to the curved surface, the section of which is indicated by a heavy line. The volume ΔV that is surrounded between the plane YY' and the curved surface represents the transport-volume during Δt, and $\Delta V/\Delta t$ becomes a measure that will stand for the intensity of the protoplasmic streaming. The form of the curved surface as well as the diameter of the stream is constantly changing as time passes. Therefore it is very impractical to follow the change in intensity $\Delta V/\Delta t$ by mapping the velocity distribution successively.

This value can be obtained, however, simply by measuring the inclination of the transport-volume curve against the time axis. The lower wave train in Fig. 21 is an example of this. In Fig. 21 (bottom), it is seen how $\Delta V/\Delta t$, or the intensity of the protoplasmic streaming, changes with time. The moment in which the streaming direction is reversed corresponds in

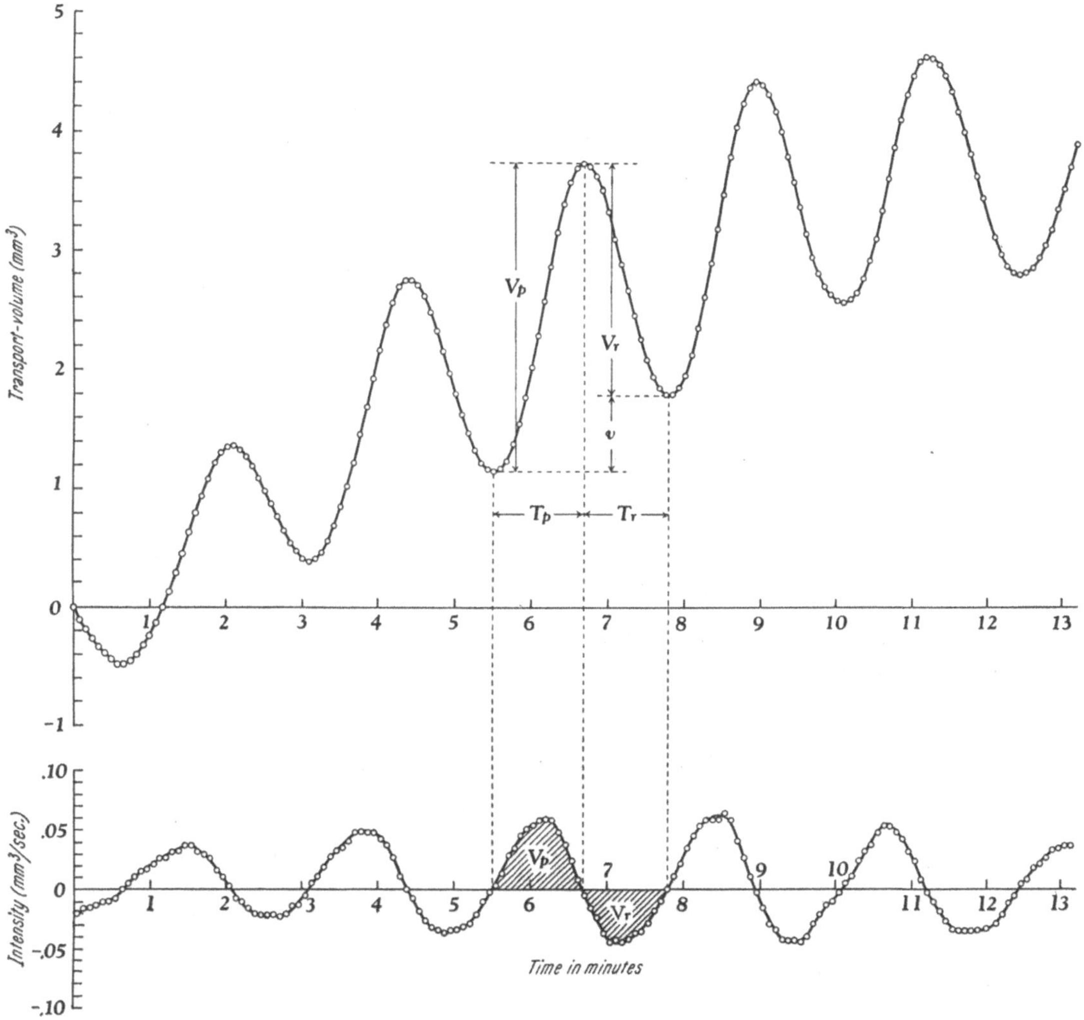

Fig. 21. Top: curve showing the transport-volume of protoplasm. Bottom: curve showing the intensity of protoplasmic flow. V_p, V_r: volumes of protoplasm carried by a progressive and a regressive streamings having streaming durations of T_p and T_r, respectively. v: net difference between V_p and V_r. (KAMIYA 1950 c.)

Fig. 21 (bottom) to the moment at which the curve crosses the time axis, and the transport-volume of the protoplasm within a certain definite period of time is represented by the area that is included between the curve and the time axis. The maximum intensity of the intracapillary flow reaches 6×10^{-2} μl./sec. in Fig. 21.

Thus through a procedure of differentiation by graphical method, a curve is elaborated which shows the intensity of the protoplasmic flow in the myxomycete plasmodium.

IV. Motive Force Responsible for the Protoplasmic Streaming

It is a relatively simple matter to express the protoplasmic streaming by means of the speed of flow. But the rate of the streaming is a function not only of the force behind the streaming, but also of the viscosity of protoplasm, which is likely to change in accordance with internal and external conditions. Consequently even if we recognize the change in speed of flow under some external influence, we cannot determine just how far the change is due to the increase or decrease of the motive force and how much of it is due to changes in viscosity (cf. HEILBRUNN 1952, p. 366). In case of a plasmodium the ever-changing width of the channel (diameter of the capillary) along which endoplasm streams is another factor which exerts an influence on the streaming rate. Physiological studies on protoplasmic streaming, which have hitherto taken almost exclusively the speed of flow as a criterion of the activity, seem to have been hampered in their progress at this point. If we want to go still closer to the causes in our understanding of the phenomenon, it becomes essential to overcome the technical difficulties in measuring the motive force of the streaming.

1. Motive force of the rotational streaming

It is only quite recently that the absolute magnitude of the power of protoplasm to cause rotational streaming has been obtained. The experiment was done by KAMIYA and KURODA (1958 b) using *Nitella flexilis.*

a) Principle of measurement

The first thing necessary for calculating the absolute magnitude of the motive force of rotation is to know where the motive force is generated and how it is exerted. Through the analysis of velocity distribution of the protoplasmic streaming in *Nitella* (p. 19 and p. 139), KAMIYA and KURODA (1956) arrived at the conclusion that the seat of the driving mechanism is located at the interface between cortical plasmagel and inner plasmasol. Independent of the as yet unknown nature of the energy that generates the motive force, it has been shown that this force is exercised in such a manner that the outer edge of the endoplasmic sol layer is shifted along the inner surface of the cortical gel layer. Accordingly, the magnitude of the motive force for this type of streaming is expressed by the shifting force that is generated by unit area of sol-gel interface. Its dimension is therefore expressed by $[ML^{-1} T^{-2}]$, that is, the same as pressure, but it differs from the pressure in that the force is exerted in parallel to the interface.

When an internodal cell is centrifuged moderately in its longitudinal axis, we notice that the streaming toward the centrifugal end of the cell is accelerated while the streaming toward the centripetal end of the cell is retarded. If the centrifugal acceleration is increased beyond a certain value, the endoplasm moving toward the centripetal end of the cell is forced backward. VIRGIN (1949) found in the *Elodea* cell that the chloroplasts collected at the centrifugal end of the cell start to go back centripetally while the centrifuge is still running, if the centrifugal force becomes low (see p. 104). In *Nitella* it is possible to find out the speed of centrifuge rotation at which the streaming toward the centrifugal end in direct proximity of the cortical gel is brought to a standstill. This centrifugal acceleration, which was referred to as 'balance-acceleration' by KAMIYA and KURODA, varies according to the cell as well as to the thickness of the endoplasmic layer.

In order to calculate the motive force from the balance-acceleration, however, we must be further acquainted with the difference in density between the cell sap and the endoplasm as well as the thickness of the endoplasmic layer. KAMIYA and KURODA presented the following formula for calculating the motive force of the protoplasmic rotation:

$$F = A\,d\,(D_p - D_s)\,a$$

where F represents the motive force, *i. e.*, the shifting force at the boundary, a, the balance-acceleration, d, the thickness of the endoplasmic layer under balance-acceleration, D_p and D_s, densities of endoplasm and cell sap respectively, and A, the area of the interface between the cortical gel and endoplasmic sol layer, where the motive force is produced.

b) Experimental data

(1) *Densities of endoplasm and cell sap* (D_p, D_s). The density of endoplasm of *Nitella flexilis* was known by KAMIYA and KURODA (1957 c) to be 1.015 gm./c.c. by the gradient method and 1.014 gm./c.c. by the falling sphere method. The density of the cell sap was found by observing whether the cell sap went upward or downward when the cell was placed with its long axis vertical in the isotonic media of known densities and was cut at its both ends. It was found to be between 1.010 and 1.011. Consequently the difference in densities between endoplasm and cell sap may be regarded as 0.004.

(2) *Thickness of the endoplasmic layer* (d). The thickness of the layer of the endoplasm when it is at a standstill under balance-acceleration must be measured directly under the centrifuge-microscope.

(3) *Balance-acceleration* (a). The centrifugal acceleration which is just strong enough to stagnate the centripetal flow of endoplasm depends upon the thickness of the endoplasmic layer. It is noticed, especially clearly in the rhizoid cell, that the endoplasmic layer often glides down under the effect of centrifugal acceleration over the cortical gel layer as a whole without causing appreciable shear inside (cf. p. 106). But, no matter whether shear is caused or not inside the endoplasm, what is needed now is the centrifugal acceleration which just stops the centripetal endoplasmic

streaming in the direct proximity of the cortical gel. It is usually in the range of 200 × g–800 × g (KAMIYA and KURODA 1958 b).

KAMIYA and KURODA (1958 b) present the result of their preliminary experiment in Table 2.

We see in Table 2 that the motive forces coincide pretty well with each other independently of the thickness of the endoplasmic layer. We thus know that the motive force generated in the internodal cell of *Nitella flexilis* is within a range as low as 1–2 dynes/cm.^{2} at room temperature.

Table 2. *Data for the calculation of the motive force of the protoplasmic rotation in internodal cells of Nitella flexilis.*

Experiment No.	Temp.	Balance-acceleration (α)	Thickness of the endoplasmic layer while it is at a stand-still under balance-acceleration (d)	Motive force per unit area (F/A)
	C°.	× g	μ	dynes/cm^2.
1	27	660	4.5	1.2
2	27	510	8.5	1.7
3	27	400	11.5	1.8
4	27	360	13.0	1.8
				mean = 1.6

So far these are the only experimental data which give the absolute strength of the force causing protoplasmic rotation, and further studies along this line are awaited.

2. Motive force of the protoplasmic streaming in the myxomycete plasmodium

It is in the slime mould that the motive force of the streaming has been so far most exactly mesured and its physiological properties most thoroughly studied. In this case, as will be stated in a later chapter (p. 142), the streaming is brought about by local difference in the inner pressure of one and the same plasmodium. Therefore the motive force is expressed in terms of counter-pressure which is just sufficient to stop the streaming.

a) Outline of the double-chamber method

For measuring the motive force of the streaming in the myxomycete plasmodium KAMIYA (1940, 1942, 1953) used a special chamber divided into two compartments. In each of the compartments there is a small mass of protoplasm connected with the other by a delicate strand of protoplasm penetrating the central septum (Fig. 22). The construction is such that the two compartments are kept air-tight without the septum separating them blocking the protoplasmic flow in the connecting strand. While one (*A* in Fig. 22) of the two compartments is kept at atmospheric pressure, the air pressure in the other compartment (*B* in Fig. 22) is under control. The endoplasm flows back and forth between the two protoplasmic masses in the different compartments along the connecting strand which is a sort of capillary with the wall made of plasmagel.

When there is no difference in pressure between compartments *A* and *B*, streaming takes place normally along the connecting strand of the more or less dumbbell-shaped plasmodium causing a corresponding change in the volume of the two protoplasmic masses. This streaming shows the usual rhythmic reversal in the direction of flow which is so characteristic of the slime mould. When, however, pressure difference is once established between the two compartements, the streaming is strikingly affected. It is

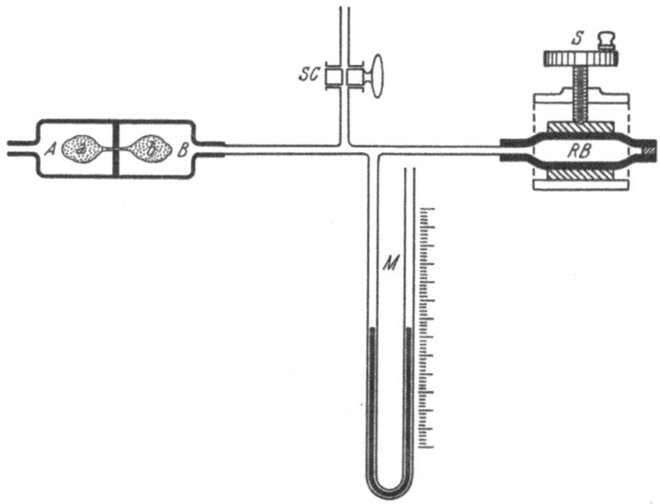

Fig. 22. Diagram showing the general arrangement for measuring the motive force of protoplasmic streaming in a myxomycete plasmodium. The whole system consists of a double-chamber having compartments *A* and *B*, manometer *M*, stop-cock *SC* and rubber bulb *RB* the volume of which is controlled by screw *S*. Protoplasmic mass in *A* and *B* is designated *a* and *b*, respectively.

either accelerated or retarded, or reversed according to the direction and magnitude of the pressure difference applied.

By making the air pressure in compartment *B* either higher or lower than that in compartment *A*, it is possible to oppose the streaming in such a way as to hold the endoplasm at a standstill. The counter-pressure just sufficient to prevent the protoplasm from flowing either forward or backward, is the compensation pressure of the power of protoplasm to flow, and hence is equivalent to the absolute value of the motive force responsible for the streaming. This counter-pressure has been termed "balance-pressure" (Kamiya 1940, 1942, 1953) and the method has been called "double-chamber method."

Obtaining the balance-pressure means measuring the motive force of the protoplasmic flow. The movement of the endoplasm reacts very sensitively to pressure differences so that even a slight deviation of less than 0.2 cm. of water column from the point of equilibrium will make the protoplasm start moving in either direction.

The balance-pressure can be determined independently of the viscosity of protoplasm. What is dependent on the viscosity is the sensitivity of the flow passively induced by the pressure difference established artificially between the two compartments. This situation is somewhat comparable to the situation when unknown potential difference is measured by means of a potentiometer in that the unknown potential difference is measured independently of the resistance of the circuit of a null pointer (galvanometer). What serves as a null pointer in measuring the motive force of the protoplasmic flow is the endoplasm of the connecting strand penetrating the central septum of the double chamber.

Restraining the movement by the application of balance-pressure does not seem to do any harm to the protoplasm, for even in a preparation where the streaming has been made to stop for several hours, normal movement is immediately resumed the moment the balance-pressure is eliminated.

The magnitude of the balance-pressure, and hence that of the motive force itself, changes according to a pattern of endogenous rhythm. The maximum value of the balance-pressure hitherto encountered was 40 cm. of water column, but usually the motive force rises and falls within the range of 20 cm. both on positive and negative sides.

According to OHTA (1952 a) there is no direct relationship between the magnitude of the motive force and the amount of protoplasm placed in the two compartments. This situation is reminiscent of the situation that the magnitude of the electromotive force is independent of the number of batteries as far as they are connected in parallel. Increasing the amount of protoplasm may correspond, as it were, to the parallel connection rather than to the serial connection. Of special interest and importance is the mode of the autonomic change of the motive force which will be given in the following section.

b) The curve representing autonomous changes in the motive force — "dynamoplasmogram (DPG)"

Since the motive force developed in the plasmodium changes spontaneously, the counter-pressure must be adjusted accordingly, if the protoplasm is to be kept immobile. Increase and decrease of the balancing counter-pressure, or balance-pressure, reveal autonomic variations of the force developed in the vital process.

In order to determine in what manner and to what extent the motive force changes in relation to time, instantaneous values of the balance-pressure were read at short intervals. By plotting a series of these successive values as ordinates against time as abscissas, undulating curves may be obtained which faithfully portray the distinguishing features of the change which the motive force undergoes in accordance with the endogenous rhythm of the protoplasm. The balance-pressure can also be registered continuously by the aid of a self-recording manometer.

In view of the fact that such a curve showing the change against time of the force developed in protoplasm is highly significant as a criterion

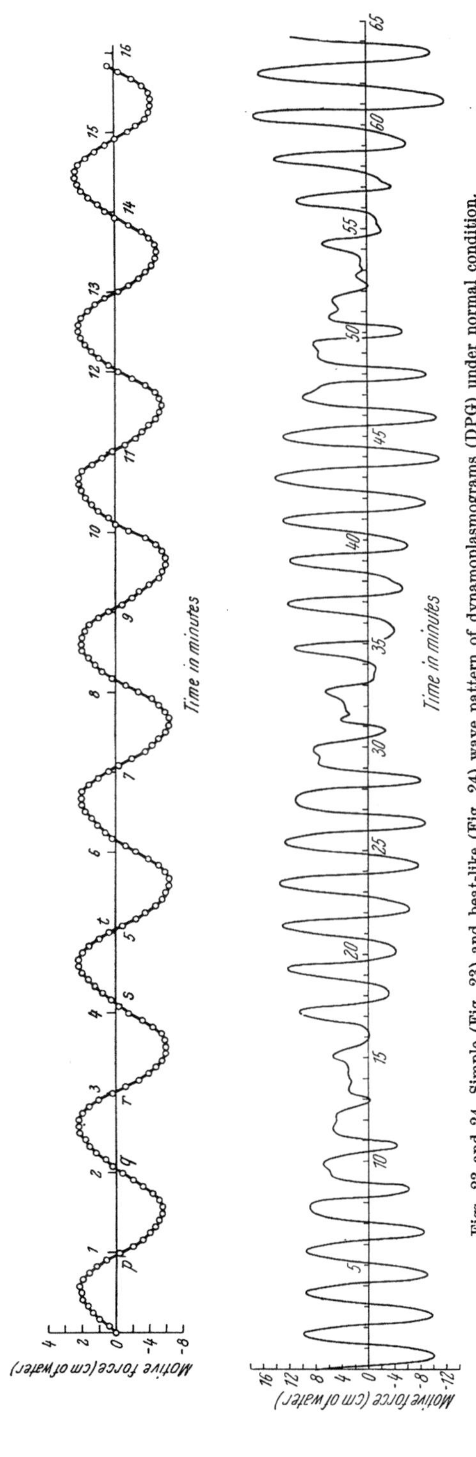

Figs. 23 and 24. Simple (Fig. 23) and beat-like (Fig. 24) wave pattern of dynamoplasmograms (DPG) under normal condition. (KAMIYA 1943, 1953.)

representing the physiological reaction of protoplasm and also in the study of rhythmicity in the living organism, KAMIYA (1942, 1943, 1953) referred to the curve by the name of "dynamoplasmogram," which will be abbreviated as DPG. All the characteristics of rhythmic activity such as wave form, frequency, amplitude and polarity are expressed by graphical representation.

The patterns of DPGs are, like those of the curves showing the rate of flow (p. 29) or the transport-volume (p. 36) different according to the specimens used and vary with time even in the same specimen. The DPG is also greatly influenced by external conditions. That the values measured fall satisfactorily into a smooth curve is an indication of the precision of measurements which have been secured thanks to the fact that the intracapillary flow reacts so sensitively to uneven pressure. It is due to this characteristic that it has been possible to find the exact value of the balance-pressure.

Fig. 23 shows a typically regular case that is occasionally met with. The curve consists of nearly perfect sine waves having a period of about 2 minutes and an amplitude of 2.5 cm. When the curve is above the base-line, the motive force works from *a* to *b* (cf. Fig. 22), while when the curve remains beneath the base-line, the motive force works in the direction from *b* to *a*. The points where the wave intersects the base-line represent the moment at which the motive force became zero; if the protoplasm were free to flow, it would correspond to the moment when the direction of the flow reverses. The streaming durations, *i. e.*, the durations in which streaming takes place in one direction (VOUK 1910), correspond on the graph to the durations represented by *pq*, *qr*, *rs*, *st*, ... and the rhythm durations by

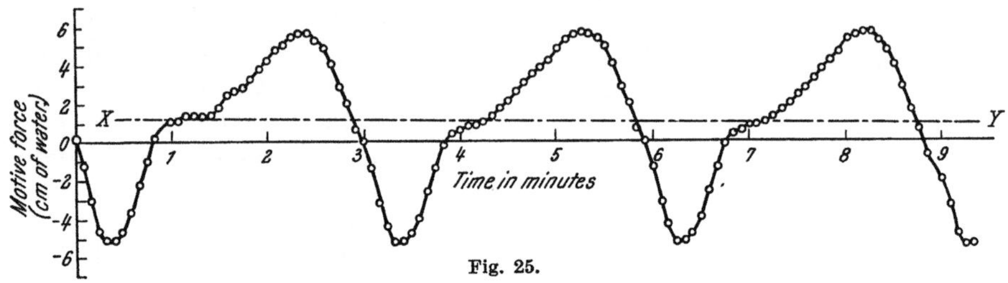

Fig. 25.

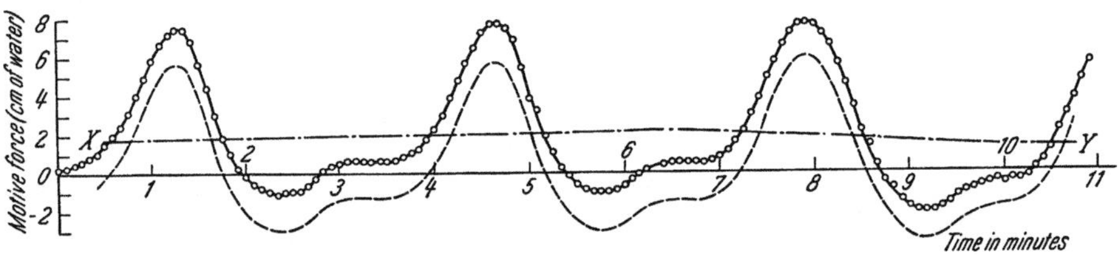

Fig. 26.

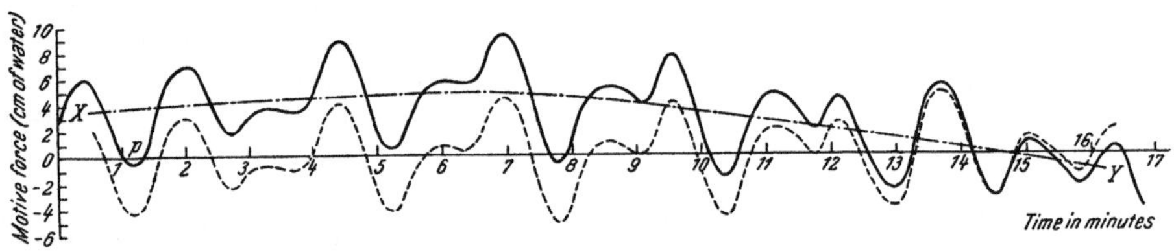

Fig. 27.

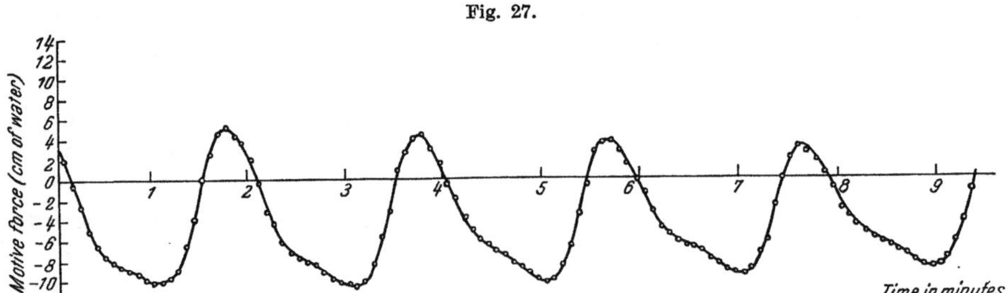

Fig. 28.

Figs. 25–28. Reiterated appearance of the characteristic wave form of dynamoplasmograms (DPG) under normal conditions.
(KAMIYA 1942, 1943.)

which are meant the sum of the durations of both forward and the following backward streamings, correspond to *pr*, *qs*, *rt*, ...

In general, the streaming duration as well as the rhythm duration and the magnitude of the motive force are different in different specimens under the same external conditions.

We encounter a great many varieties of patterns of DPGs which normal plasmodia manifest. It is interesting to note that change in amplitude often takes place with considerable regularity, *i. e.*, after a decreased amplitude phase the wave swells out gradually to a maximum, and after the lapse of a certain period of time, the magnitude decreases again and is followed by the next increased amplitude phase. A regular alternation of the amplitude is clearly illustrated in Fig. 24. In Fig. 24 the "loop" and "node" of the wave appear alternately at an interval of about 20 minutes. These reiterated turns consist each of 8–9 waves, the average wave length being about 120 seconds. This is, however, not a general rule; more examples of beat-like wave patterns were presented elsewhere (KAMIYA 1942, 1943, see also Figs. 29 and 32).

We have had ample examples of DPGs which show that the motive force has undergone pronounced variations not only in its magnitude, but also in its wave form during the rhythmic succession of vital processes. Sometimes a characteristic form appears repeatedly throughout several waves. Fig. 25 is one example of this. The three waves showing a similar pattern with a period averaging 180 seconds have maxima of 5.6 cm. and minima of — 5.2 cm. of hydrostatic head. As the curve repeats almost the same details of the changes, it may be assumed to be periodic. The dash-dot line *XY*, which will be considered in the following section, cuts the wave equally into an upper and lower halves. In Fig. 26 the wave form is also characteristic. A similar form is repeated here with a period of about 200 seconds. As for the curve represented with a broken line we shall refer to it in a later chapter.

In Fig. 27 is observed the process through which a particular wave form possessing two maxima appears in a gradually changing form in the succeeding waves at an interval of about 150 seconds. The wave train which remained above the time axis is gradually descending as a whole. The waves presented in Fig. 28 rise rapidly, then fall rather rapidly at first and later more gradually until the trough is reached, but the magnitude of the wave gradually decreases in this case. The distance between any two adjacent maxima or minima of the waves is nearly constant for about 117 sec.

c) Polarity of the motive force

In all the foregoing graphs which are but few of the selected parts of a number of the motive force curves in plasmodia under normal states, it is noticeable that the area included within the curve and the base-line on the + side differs from that on the — side. This difference between the + and — sides indicates that the motive force is not exerted equally in both directions but more in one direction than in the other. Because of this polar nature of the motive force a plasmodium, were it free to move, would advance by flowing forward a little more each time and not retreating all the way back to its original position when it reverses.

Let us suppose a smooth line which cuts an undulating curve in such a way that the area on the upper side of the line is equal to that on the lower side. If the curve is periodic, or almost periodic as in Figs. 25 and 26, the height of this line can be found easily with the aid of a planimeter or by any other practical method. When the wave form changes from rhythm to rhythm, as is the case in many of the DPGs with which we are concerned, the line in question may be assumed with a certain degree of approximation by means of the method described by KAMIYA (1942, 1943).

The dash-dot lines *XY* in Figs. 25, 26 and 27 are lines which divide the waves equally into upper and lower parts. These lines show the "true" axes of the waves, the ordinate of which is a measure of the *polarity* of the plasmodium in its absolute scale.

Holding polarity and rhythm in protoplasm as factors independent of each other, KAMIYA (1942, 1943, 1953) called the line "polarity-line." The strength of the tendency of the plasmodium to advance in one direction, *i. e.*, the strength of its taxis, has no relation to the amplitude of the waves. It is utterly under the control of polarity.

Those who have once observed protoplasmic streaming in a plasmodium will recall the "abnormal," but not unusual phenomenon that the protoplasm shows when it continues to flow only in one direction for an excessively long period of time. From such an "abnormal" flow, one would be led to believe that protoplasmic rhythm is often disturbed. If one bases one's judgement of the rhythm of protoplasm merely upon the reversal of the flowing direction, then one will overlook the following important situation.

Fig. 27 shows a case in which the whole undulating curve is on the + side of the base-line more than 6½ minutes without crossing the base-line. This means that the motive force is exerted always in the same direction during that time. Nevertheless, the undulating curve manifestly reveals a rhythmic change throughout that time.

Obtaining the polarity-line is of course necessary in the quantitative study of polarity itself, but it is also needed for eliminating non-periodic factors that might be involved in the analysis of periodic phenomena. The broken line curves in Figs. 26 and 27, which will be taken up later again, represent the measured curves from which the polarity effect has been eliminated.

At all events, it is clear from the above facts that, by merely measuring the time of reversal in the streaming direction, as was hitherto done, only one aspect of the periodic phenomenon can be observed. A complete time course of the phenomenon is indispensable in investigating the periodicity of protoplasmic streaming in a slime mould.

d) General characteristics of the dynamoplasmogram of the plasmodium

Summing up various cases as to the form of the DPGs obtained in normal, untreated specimes, it can be generally stated as follows:

1. Though there are cases presenting an almost perfect sinusoidal curve (Fig. 23) the majority of them deviates more or less from it.

2. In a constant environment, both the amplitude and the wave form

change spontaneously. Although there are some few cases where they present a fair constancy during a certain period of time (Figs. 25 and 26) both amplitude and wave form will eventually change when the measurement is continued longer.

3. The change in the wave form and amplitude are always gradual (Figs. 27 and 28), there being no cases of abrupt changes occurring without any connection with the preceding and succeeding waves.

4. Many as are the cases in which the amplitude increases or decreases irregularly, there are also not a few cases in which it alternates its rise and fall with a nearly constant period like a beat wave (Figs. 24, 29, and 32).

5. The time course of the motive force generation is never expressed in a particular wave form, but presents a form that is rich in variability both within the same specimen and among different specimens.

6. It may be assumed that polarity changes independently of the rhythmic phenomenon.

In viewing the foregoing graphs, it is to be noted that the curves show no sign of special deflection when they cross the base-line; in other words, the points which correspond to the moments of reversal are not characterized by the shape of the curve. Previous investigators have almost implicitly adopted the moment of reversal of flow as the only criterion for distinguishing one rhythm from another; for other than this, there has been no gauge with which to measure the period of the rhythm. It is evident that this method of separating one rhythm from another is most practical, but one must keep aware of the fact that the instants of reversal, where the motive force is zero, have neither a special meaning in rhythmic activity nor do they necessarily imply the same phase of rhythm as manifested by the graph. A reversal moment simply means a momentary equilibrium of the intraplasmic forces. In a later chapter we shall look into the causes through which the variability in the wave pattern and amplitude is brought about.

e) Simultaneous measurement of the motive force and the potential difference

Generation of the motive force is accompanied with changes in bioelectric potential difference. In order to find a possible relation between the force responsible for the streaming and the potential difference between the anterior and posterior parts of the flow, Kamiya and Abe (1950) measured both balance-pressure and potential difference simultaneously by combining the double-chamber method with potentiometry. The method employed was similar to that shown in Fig. 22 except that protoplasmic masses in the two compartments are led electrically to the potentiometric circuit.

By plotting on a graph two series of values concerning the balance-pressure and the potential difference one can express simultaneously changes in the force causing the streaming and the potential difference in the plasmodium in the two curves. In Fig. 29 the potential difference is designated as plus when the potential of *b* is higher than that of *a* (cf.

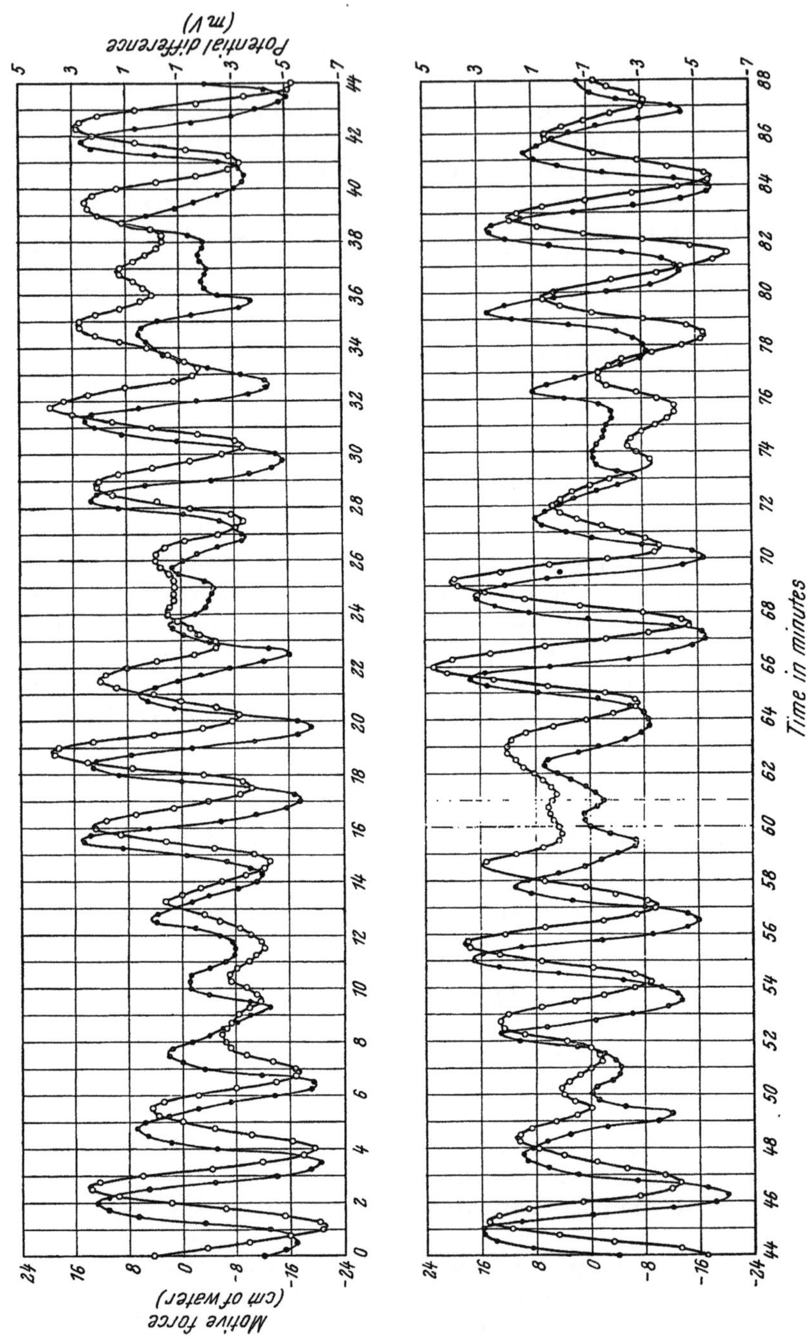

Fig. 29. Superimposed representation of the beat-like waves of the motive force and the potential difference. Motive force is shown by black circles and potential difference by open circles.
(KAMIYA and ABE 1950.)

Fig. 22), and as minus in the reverse situation. The balance-pressure is designated according to the previous custom with a plus sign when the pressure in compartment *B* is higher than that in *A*, *i. e.*, when the protoplasm tends to flow from *a* to *b*, and with a minus sign for the opposite state. Kamiya and Abe referred to such a curve representing the potential difference in a protoplasmic system as "electroplasmogram" (EPG) in analogy to the "dynamoplasmogram" (DPG).

Fig. 29 is one example showing the result of the simultaneous measurement of the potential difference and the motive force under normal conditions. This is an example of beat-like waves. As is shown in the figure, when the amplitude of DPG increases, that of the EPG increases correspondingly. We find that there are close correlations between the two waves both in regard to the wave form and amplitude. It must not be overlooked, however, that the phase of EPG always lags behind that of the DPG by about a quarter of a period. We shall return to this problem later (p. 159).

f) Simultaneous measurement of the motive force and respiration

A certain reagent applied to the plasmodium affects the generation of the motive force as well as respiration. Their reactivity is naturally different according to the reagent used, but it is also true that plasmodia behave not always constantly under the effect of one and the same reagent in the same concentration. Difference in reactivity according to individual specimens is by no means rare in biological materials and the plasmodium is not an exception. And this is why simultaneous measurement of the motive force and respiration in one and the same plasmodium is desired for studying the energetics of the motive force.

For the purpose of measuring both the motive force developed in the plasmodium (*Physarum polycephalum*) and the respiration of the same plasmodium simultaneously, Kamiya, Abe and Nakajima (1957 a) set up an apparatus the diagram of which is shown in Fig. 30. In contrast to the foregoing set-up (Fig. 22) the whole system here is closed from the outside. The pressure difference between the right and left halves of the apparatus is under control by means of the pressure controller *PC* consisting of two injection syringes of equal size combined back to back. It is thus possible with this closed push-pull system to measure the motive force of the streaming of protoplasm in the double-chamber.

When pressure difference is established between the left and right halves of the apparatus, the oil droplets D_L and D_R in the capillaries naturally shift both toward the left or both toward the right according to the direction of the induced difference in pressure. In theoretical treatment, the condition is known under which the distances of shift covered by the two drops in the capillaries remain equal to each other under an arbitrary pressure difference. That condition is:

$$\frac{v_L}{V_L + v_L} = \frac{v_R}{V_R + v_R}$$

where v_L and v_R are the gas spaces of the compartments A_L and A_R respectively including the spaces of capillaries up to the oil drops, and V_L and V_R are the remaining gas spaces on the left and right sides of the apparatus beyond the oil drops D_L and D_R.

Under this condition the sum of the gas spaces of the two compartments, $v_L + v_R$, remains the same no matter how the pressure difference is set up

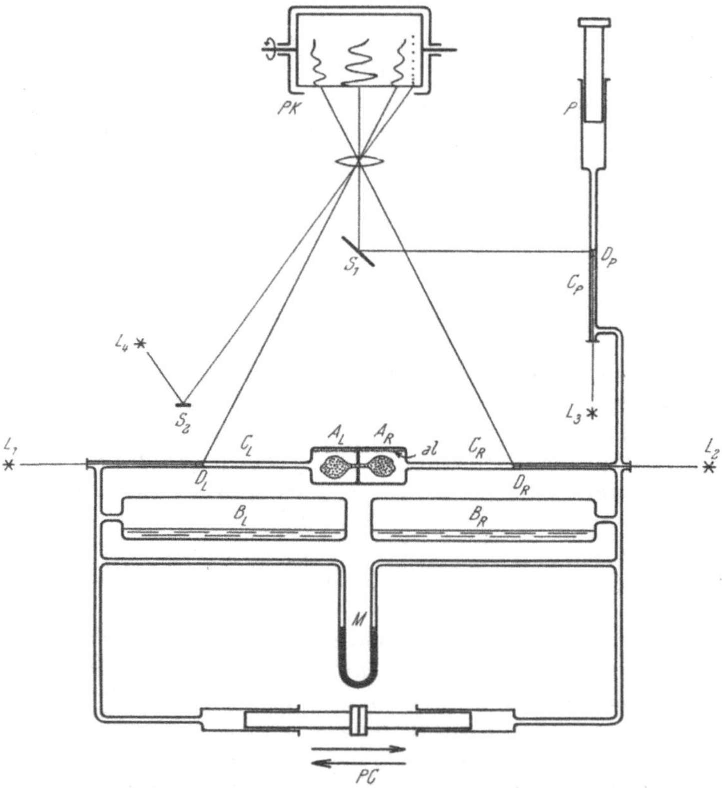

Fig. 30. Diagram of the apparatus. A_L, A_R: compartments of the double-chamber containing the material and the pieces of filter paper *al* soaked with alkali. B_L, B_R: compensation chambers partially filled with water. C_L, C_R: glass capillaries containing kerosene droplets, D_L, D_R. *PC*: pressure controller. D_P: pressure indicator in the glass capillary C_P. *M*: manometer on which D_P is calibrated as an indicator of the pressure difference between A_L and A_R. *P*: injection syringe for adjusting the sensitivity of D_P to the pressure. L_1, L_2, L_3: light sources for illuminating the menisci of the index droplets. L_4: light signal with an interval of 1 min. *PK*: photokymograph.
(KAMIYA, ABE and NAKAJIMA 1957 a.)

between the left and right. Thus if we regard $(v_L + v_R)$ as the gas space of just one reaction chamber and $(V_L + V_R)$ as that of one compensation chamber, the whole system is comparable to a simple capillary volumetric respirometer of a differential type.

In order to measure the motive force of the streaming continuously the endoplasm in the connecting strand must be kept at a standstill by adjusting the pressure difference between the left and right without interruption. This procedure, however, does not modify the distance between the two index droplets, D_L and D_R, since both of them move in parallel fashion in the capillary. The distance between them changes only when the gas is liberated or absorbed in the compartments containing the material. Therefore, if a small piece of filter paper soaked with alkali is placed both in compartments A_L and A_R, CO_2 gas liberated from the plasmodium is absorbed and the distance of the two drops becomes gradually smaller as oxygen is consumed by the plasmodium. The rate

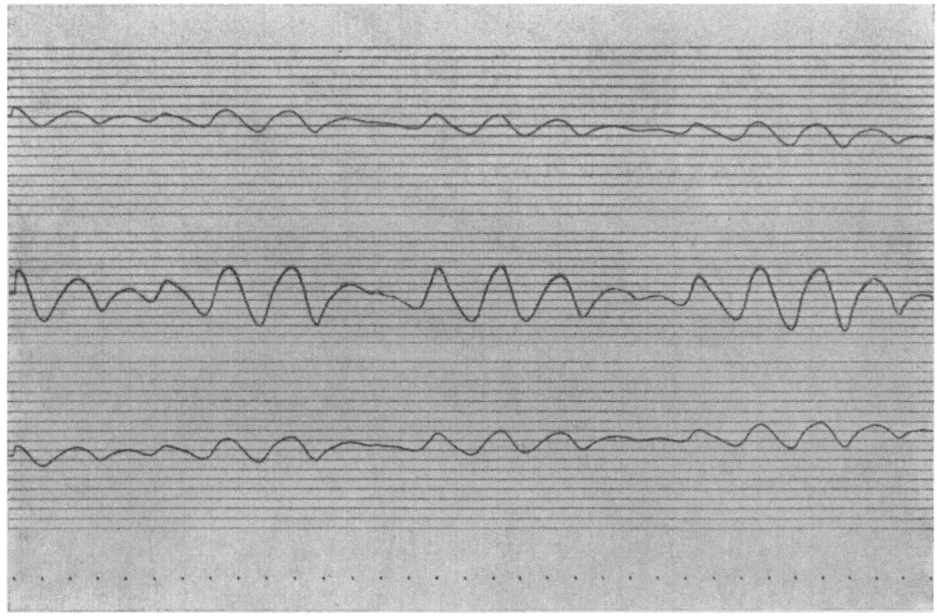

Fig. 31. A photographic record obtained from normal material. The central wave: DPG. [The unit interval of parallel striations in the middle shows a pressure difference of 4 cm. of water column between A_L and A_R. The initial position of the curve at the left end shows the zero point where air pressures in A_L and A_R are equal.] The oxygen uptake is indicated by the gradual decrease in the distance between the top and bottom waves. [The unit interval of parallel striations here corresponds to 10 μl. of oxygen uptake.]. Time interval: 1 min. (KAMIYA, ABE and NAKAJIMA 1957a.)

of oxygen consumption can be calculated from the rate of approach of the two drops.

For the practice of the simultaneous measurement of the motive force and respiration with this apparatus, it is necessary to determine the pressure difference just sufficient to keep the endoplasm in the connecting strand at a standstill, as is the case with the ordinary dynamoplasmometry mentioned in a foregoing section and, at the same time, the distance between the two index droplets. In order to perform such a measurement conveniently and continuously without personal error of observation, KAMIYA, ABE and NAKAJIMA resorted to photographic registration. Adjustment of the pressure difference between A_L and A_R which is just enough to stop the flow has been conducted by the pressure controller PC, always keeping watch on the protoplasm in the connecting strand through the inverted microscope. The pressure difference referred to is known from the manometer M, but is also represented in this apparatus by the position of the droplet D_P. A photographic record of the position of the droplets D_L, D_R and D_P is shown in Fig. 31.

By plotting the measure of approach of the two index drops, D_L and D_R, as ordinate against time as abscissa, a straight line is obtained which shows

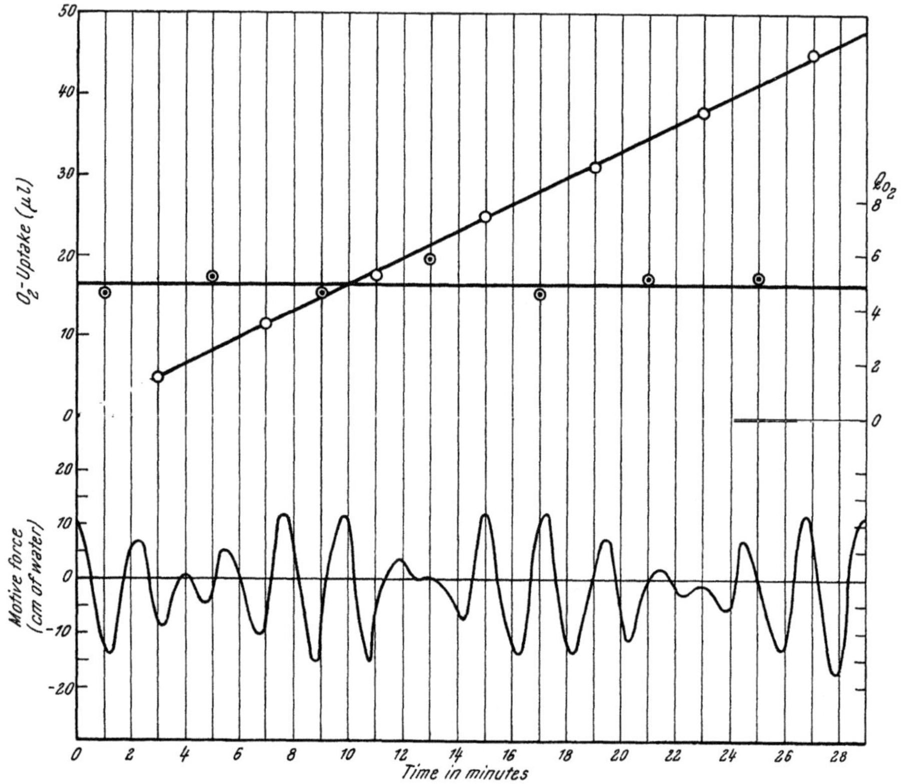

Fig. 32. Simultaneous representation of the motive force (bottom), oxygen uptake (circles) and Q_{O_2} (μl./mg. dry weight/hr.) (figures with dots) in one and the same plasmodium. (KAMIYA, ABE and NAKAJIMA 1957 a.)

constancy in the rate of the oxygen uptake (Fig. 32). Q_{O2} was calculated to be 5.0 on a dry weight basis in this case, which corresponds satisfactorily to the results obtained with the Warburg manometer by previous authors (ALLEN and PRICE 1950, OHTA 1954). Attention should be paid to the fact revealed from Fig. 32 that the oxygen consumption proceeds at a constant rate while the amplitude of the wave changes conspicuously. We can therefore say that phases of increased and decreased amplitude of the DPG do not represent the increased and decreased phases of respiration. Further results obtained with this apparatus will be introduced in Chapter VI.

V. Rhythmicity

In protoplasmic streaming there are cases when no rhythmicity whatsoever is observed in the movement as in the typical rotation seen in *Characeae,* while in other cases the visible movement manifests conspicuous rhythmicity as is the case in myxomycete plasmodium. As to circulation, its movement has a rhythmical nature in that the streaming direction reverses itself from time to time. KÜSTER (1956) claims that one of the most important points distinguishing rotation from circulation is that the former is aperiodic while the latter is periodic. Without doubt the most remarkable and regular rhythm manifested by the protoplasmic streaming is found in myxomycete plasmodia. There is a classical work of VOUK (1910) on the rhythmicity of the streaming in the plasmodium of *Didymium nigripes.* He measured successively the duration of forward and backward flow and came to the conclusion that the total duration of progressive and regressive streamings, or what is called "rhythm duration" by VOUK, is constant.

1. Rhythm of motive force generation in the plasmodium

It has already been shown in the foregoing how the protoplasmic streaming in the plasmodium of *Physarum polycephalum* changes its rate and transport-volume spontaneously. The curves representing them were one and all closely correlated to a motive force curve, but they deal with more complicated quantities than the motive force itself, since both the transport-volume and rate of the streaming are functions not only of the motive force, but also of various factors which are not kept constant during the rhythmic flow, such as width of the streaming channel, viscosity of endoplasm, etc. Moreover, the changes in the distribution of protoplasm which are necessarily brought about by allowing free movement will change in turn the inner conditions generating the motive force. Therefore, the analysis of whatever curve is obtained while the plasmodium is left free to move is rendered more difficult and complicated than that of the motive force causing the flow.

The measurement of the motive force, or dynamoplasmometry by means of the double-chamber method, is based on the principle of compensation. The balance-pressure can be determined independently of the viscosity of protoplasm and of the inner diameter of the capillary and because of the balance-pressure there is no change in the volume of the blobs of protoplasm in the two compartments during the experiment. Thus it can be said that the DPG obtained through the compensation method expresses in a far purer form the aspects of protoplasmic activities as compared with the record of any visible motion.

a) Analysis of dynamoplasmograms

We are now confronted with the problem concerning the nature of the rhythm operative in the protoplasmic system. Leaving aside the question of polarity, we shall first consider the fact that the amplitude of a DPG obtained under constant conditions increases and decreases entirely according to an autonomic scheme. If we assume that the periodic rise and fall of the amplitude of the DPG is due to the "physiological disturbance" or "fatigue" of protoplasm, then we shall have to see just what is meant by physiological disturbance or fatigue. A further question is the cause of the repeated appearance of a characteristic wave form in successive rhythms.

In order to analyse and understand the meaning of a non-simple oscillatory phenomenon, it is often significant to break it down into a group of simple oscillations by means of an adequate method. This way of thinking implies that a non-simple oscillation is no other than the sum of many simple oscillations. The attempt to understand the changes in the purely biological force originating within the protoplasm through a totally unknown process by analysing them into a group of sinusoidal waves is of course justified only on the assumption that the basic mechanism develops the motive force as a sine function of time. Though the adequacy of this premise may still be open to discussion, the facts that 1. the DPGs never manifest themselves as particular curves with certain definite characteristics, 2. the changes in the wave form and amplitude are always transitional and 3. the DPGs themselves often present a sine form that is adequate enough, lend objective support to the assumption that the basic formula of the motive force generation is expressed, at least in a first approximation, as a simple harmonic function of time.

As a result of the study of many DPGs, KAMIYA (1942, 1943, 1953) arrived at the view that fluctuability of the amplitude and variability of the wave form can be most reasonably and easily interpreted under a unified conception if we assume that they are the result of the summation of many sine waves having different periods and amplitudes. The various wave forms obtained as DPGs are thus to be considered as the results of the superposition of many sine waves. On the basis of this assumption we shall try to consider the various cases of the DPG patterns in terms of simple harmonic oscillations.

The synthesis of many simple harmonic oscillations into one movement is often seen among the phenomena observed in physics, but an actual example of such a phenomenon is hardly known yet in biology. From this viewpoint too, therefore, the DPG of myxomycete plasmodia is noteworthy as an instance in biology of many oscillatory components being synthesized into one oscillation.

1. *Beat waves.* The alternating increase and decrease of the amplitude such as seen in Figs. 23, 29, and 32 bear a strong resemblance to the phenomenon of beats. In fact, increase and decrease in the amplitude of the DPG are assumed to have resulted from the interference of intraplasmic periodic forces, the frequencies of which are slightly different from each other. At the increased amplitude phases of the DPG, two main component waves reinforce each other, whereas, at the decreased amplitude phases these two main component waves oppose each other. If we assume the distance between the "nodes" as 20 minutes and the number of waves included in

it as 8, the periods of the two component waves in Fig. 23 will have a difference of 20 seconds between each other. Consequently it will be gathered that those waves have approximate periods of 140 and 160 seconds respectively. In the case of Fig. 29 the period ratio of the two main components will be about 4 : 5.

The interference of two component waves having slightly different frequencies is supposed to be responsible for the major changes in amplitude, but it cannot be the general principle governing the movement of the plasmodium. In order to explain the repeated appearance of a peculiar wave form throughout several successive rhythms, many subordinate components of different frequencies must be attributed to the protoplasm.

2. *Periodic waves.* Any one-valued periodic curve can be analysed on the basis of the Fourier's theorem into harmonic curves of suitable amplitudes

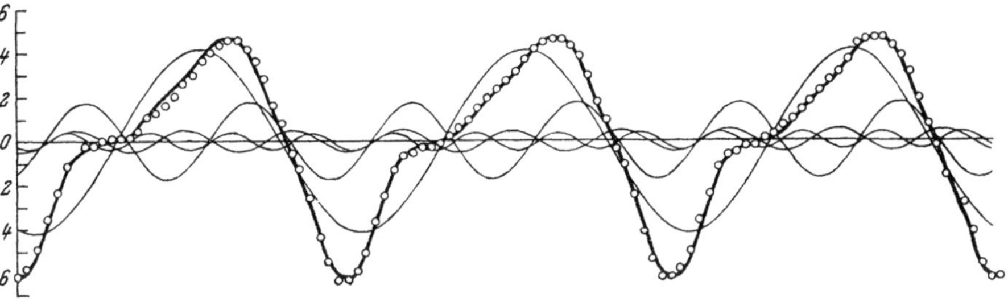

Fig. 33. Analysis of DPG in Fig. 25 into a series of harmonic curves. (KAMIYA 1943.)

and phases having wave lengths of 1, $^1/_2$, $^1/_3$ and successive aliquot parts of the wave length of the wave to be analysed. The following formula represents the result of Fourier's analysis of the DPG in Fig. 25 in terms of cosines till fourth harmonics.

$$f(t) = 4.18 \cos\left(\frac{360^0}{180}\, t - 200^0\right) + 1.70 \cos\left(2 \times \frac{360^0}{180}\, t - 150^0\right) +$$

$$+ 0.46 \cos\left(3 \times \frac{360^0}{180}\, t - 193^0\right) + 0.40 \cos\left(4 \times \frac{360^0}{180}\, t - 224^0\right).$$

The cosine coefficients represent the height of the water column in cm. When the fundamental harmonics having periods of 180 sec. and the higher harmonics are combined as in the above formula, $f(t)$ takes a form as shown in a heavy line in Fig. 33. The small circles in Fig. 33, the polarity effect being eliminated from the DPG of Fig. 25, have been plotted in order to compare how satisfactorily the rhythmic activity of protoplasm will coincide with the superposed resultant of the four components.

In Fig. 26 too, the three waves are approximately periodic when we substract the polarity effect (broken line curve). If we calculate Fourier's coefficients in regard to wave form, obtained by taking the mean of the three waves, and draw a graph by taking only the first two components, we obtain a wave form represented in Fig. 34 (heavy line), which approaches closely enough that of the broken line curve in Fig. 26. It is generally sufficient to take the first few components of Fourier's series in order to represent the periodic wave form of DPG. In such a wave form, Fourier's series rapidly converges as the number of harmonics increases.

3. *Non-periodic waves.* A special method must be resorted to in order to find out a hidden period in a case like Fig. 27 where the wave form changes little by little. If we try to analyse an arbitrary range of DPG in Fourier's series, we shall get various harmonic components which, however, do not

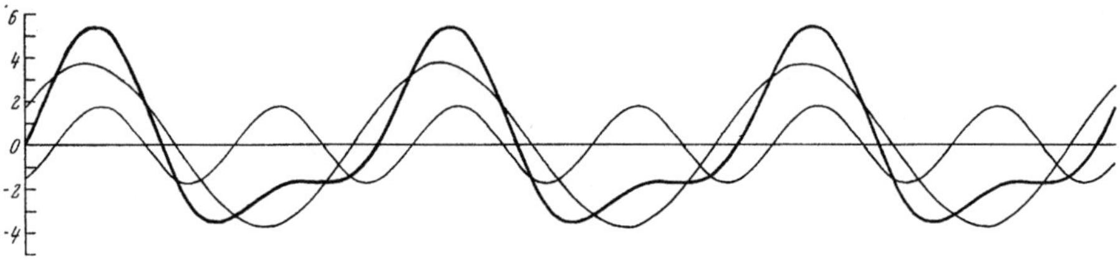

Fig. 34. The first and second harmonics obtained from DPG in Fig. 26. The heavy line is the resultant of the two harmonic curves.
(KAMIYA 1943.)

necessarily mean an entity of any physiological significance. The important thing for us is not to represent an experimental formula for the limited range of the empirical curves by means of apparent sine or cosine coefficients having various periods, but to decide which among the various periods has an amplitude that is great enough to have an independent significance. For this purpose KAMIYA (1943, 1953) applied the method of periodogram analysis and found that for the curve in Fig. 27 (broken line) there are 2 powerful components having periods of 142 sec. and 79 sec. Their

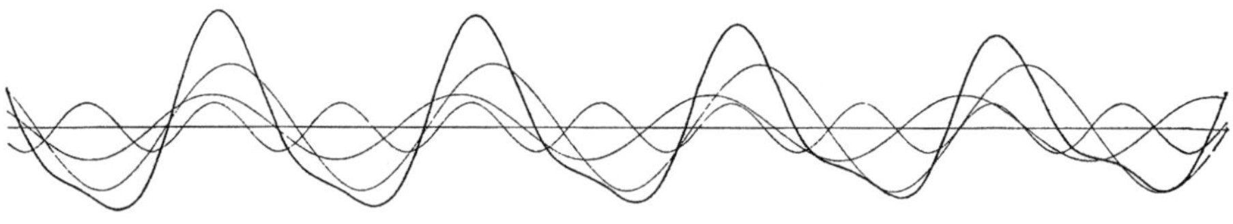

Fig. 35. Synthesis of three sine waves. The resultant shown by a heavy line bears a strong resemblance to the DPG in Fig. 28.
(KAMIYA 1942.)

amplitudes are calculated as 2.6 and 2.0 cm. of water pressure respectively. Difficult as it is to discover the periods of the hidden waves from non-periodic waves, the record of which is not long enough, we can nevertheless infer the approximate periods, amplitudes and phases of the components whose amplitude is great enough and whose periods are not too close together.

In Fig. 28 very similar wave forms are repeated, but the amplitude does not remain constant during the experiment. This curve is also non-periodic within the range under consideration. KAMIYA (1942) tried here to get information from this curve empirically; by varying the amplitude and wave lengths of the components and shifting them arbitrarily along the axis, very similar resultant waves can be reproduced from three components. Fig. 35 shows the redrafting of Fig. 28. The resemblance between the experimental curve and the resultant of the combination of three harmonic curves is satisfactory. The serial ratio of the periods of the three components is 90 : 87 : 44; *i. e.,* two of the three components have only slightly different periods, whereas the other component has a period but half as long as the other two. The peculiar asymetric pattern of the waves is due to the component of short wave length.

A point that should be raised here is that, even though a given part of a DPG is analysed into a series of harmonic curves each of which has physiological significance, it does not necessarily mean that this combination represents also the part of the DPG that is not under consideration. Even if one could analyse a limited portion of a DPG into an exact series of components, the combination of the same components would not represent the same DPG unlimitedly. An important reason why the components postulated for one portion of the curve cannot represent other portions seems to lie in the instability of the components (Kamiya 1942, 1953). It can easily be imagined, for instance, that the change in the surface-volume ratio occurring during the experiments will directly influence the metabolic activities and that this in turn will have an effect upon the rhythm of the motive force generation. Recently, Kishimoto (1958 a) tried to analyse the electric potential rhythm of the plasmodium statistically from the standpoint of the instability of the period.

In discovering and isolating hidden unknown periods from non-periodic waves by means of periodogram analysis, especially when the unknown periods are close to each other, it is necessary to increase the 'resolving power' of the periodogram by increasing as much as possible the duration of the dynamoplasmometry. The instability of the components themselves, however, sets limits to the scope of application of this method and to the exactitude of the results obtained through it, for periodogram analysis presupposes stability of the periods of the component waves.

b) Polyrhythmicity of motive force generation in the plasmodium —concept of intraplasmic interference

In the foregoing pages we have made it clear that the curve experimentally obtained can be reproduced very satisfactorily by superimposing a few sine waves. The important thing in understanding rhythmicity in protoplasm is to grasp the fact that in a protoplasmic system rhythms with different periods are working not as mere abstractions but as entities having a physiologically independent significance.

As stated before, the components are unstable; they may not remain constant during a period of time. But the components, even though they are variable, should have an identity of their own. One and the same plasmodium is most likely to produce many frequencies of rhythm simultaneously. That the DPG does not present wave forms with common characteristics, in other words, that there is always a rich variability and multiformity in the wave form, rather heightens the physical significance of a series of oscillatory components.

Seifriz (1942) observed by means of time-lapse motion picture how three independent rhythms pulsate in neighbouring parts of a plasm expanse of *Physarum polycephalum;* no two of three were pulsating in unison. In the author's personal observation of this film, these separate rhythms appeared to him as differing not only in phase but also in pulsating frequency. Such a phenomenon is highly interesting in that it renders visible to a certain extent the facts stated above, and supports the significance of the component rhythms obtained by the analysis of wave forms.

Another point to be raised here is that the individual harmonic component curves obtained through analysis are, physiologically speaking, not necessarily single vibrations originating in a single source. It is more natural to consider them as representing the results of the superimposition of sine waves, generated in a great many minute loci of the plasmodium, waves which have the same periods but are in different phases. Are the phases of the dynamic waves originating there with the same periods controlled entirely by chance on a statistical level, or is there any mechanism equalizing the phases? Harmonic analysis enables one to decompose a nonsinusoidal wave into harmonic waves, but it is utterly powerless in answering the above question. The author here recalls with interest the view of Adrian and Matthews (1934) and Motokawa (1943) on the formation and suppression of the waves in electroencephalograms (EEG). Though there seems to be apparently no connection between electroencephalogram and dynamoplasmogram, there may be unexpected common elements related to the problem of the synchronization and interference of physiological rhythms.

As a result of analysing and weighing many instances of DPG, it has been shown that there co-exist in a continuous protoplasmic system not only rhythms with nearly similar periods, but also rhythms with short periods like $^1/_2$, $^1/_3$ or even smaller fragments of the basic period. Peculiar wave forms often found in DPGs are brought about by components having short periods and gradual changes in amplitude are caused by interference among components whose periods are slightly different from each other. That is, by assuming that protoplasm is a *polyrhythmic* system, it has been shown that the complexity and multiformity of the periodic activity of the protoplasm could be unified under a general concept. However great the number of DPG patterns may be, they are only due to the difference in the mode of superimposition of simple harmonic oscillations and in part to polarity. In a word, the conclusion derived from the analysis of DPG is the remarkable fact that in a protoplasmic system there co-exists a group of physiological rhythms with various periods which are interfering with one another. Kamiya called this "intraplasmic interference."

Though it is likely that harmonic components with different periods originate in different structural elements of protoplasm, it is remarkable that components having not only slightly different frequencies, but very divergent ones are operative in the same protoplasmic system. When an elastic body having an infinite number of degrees of freedom, *e. g.,* a stretched string, vibrates, it produces a series of overtones simultaneously the periods of which are 1/2, 1/3 ... and successive aliquot parts of fundamental vibration. What is then the physical nature of the "overtones" of the protoplasmic force? Could they originate, structurally or physiologically, from a common source, just as in the case of a series of overtones from the vibration of one and the same string; or is divergency of period attributable to the structural heterogeneity of the protoplasm? This is an interesting problem which the analysis of the DPG brings forth.

c) Approaches to the study of the causes of rhythmic flow

At this point we should like to consider the cause of the periodic reversal of the streaming direction with the available experimental facts.

The implicit assumption so far made by some early workers (HILTON 1908, WATANABE, KODATI and KINOSHITA 1937) for explaining the periodic reversal of protoplasmic streaming in the plasmodium is that new conditions brought about by the streaming in either direction, such as the inner pressure which might increase at one end of the plasmodium, or any other possible physical or chemical changes brought about by the change in distribution of the endoplasm accompanying the streaming, would induce the protoplasm to flow in the opposite direction. In other words, such a view is based on the premise that the streaming is a *sine qua non* element in the chain of cause and effect governing the periodic back and forth movement. According to this view, therefore, if we could restrain the protoplasmic movement in some way or other, the motive force of the flow would not be able to advance from the phase where it stands at the time.

We know already, however, that the periodic generation of the motive force persists even when the protoplasmic flow is stopped continuously by applying balancing counter-pressure. This shows that the translocation of the endoplasm does not play a direct part in causing the reversal of the streaming direction.

This view has been ascertained still more decisively by KAMIYA and NAKAJIMA (1955) who arrested the streaming by temporal gelation. They admitted to the plasmodium of *Physarum polycephalum* CO_2 gas, which was found by SEIFRIZ (1941, 1942) to be the most satisfactory agent for gelating or 'setting' the slime mould protoplasm reversibly with the least ill effect (p. 116). When carbon dioxide gas diluted to a proper concentration is admitted to the plasmodium, the protoplasm gelates abruptly after a time. The protoplasm once gelated solates again when carbon dioxide gas is replaced with fresh air. Solation takes place even spontaneously in the carbon dioxide gas provided the gas concentration is not too high.

Placing the plasmodium in the double-chamber, KAMIYA and NAKAJIMA conducted dynamoplasmometry for some time; then the plasmodium was treated in its entirety with CO_2 diluted with air; it soon gelated. That gelation took place was indicated by the fact that no passive flow could be induced along the connecting strand even when a pressure difference was established between the two compartments. Naturally, measurement of the motive force was no longer possible under such a condition. As soon as the protoplasm solated again, measurement of the motive force could be resumed. The curve (DPG) thus obtained was composed of a part covering the normal activity before carbon dioxide was applied and of another part covering the activity after recovery.

What invites our special attention in this case is the phase relationship of the waves before and after gelation. For it was clearly shown that the waves that come after gelation are found to be just in such a phase as would suggest *a continued periodic activity during the gelated period.* If the rhythmic functioning had been stopped during the period in which the whole protoplasm was in the state of a gel, the protoplasm would have taken up after recovery the phase at which it left off just before gelation. But actually this was not the case. This fact gives sufficient grounds for reasoning that the phase of the periodic activity generating the motive force was kept advancing while the protoplasmic flow was at a complete standstill. In other words, the mechanism controlling the periodic generation of the motive force must have been functioning no matter whether the protoplasm was free or arrested, no matter whether it was in a state of sol or gel.

A further proof of this conclusion is to be drawn by figuring out the electric potential rhythm when gelation takes place all over the plasmodium. By applying the technique of KAMIYA and ABE (1950) for simultaneous measurement of the motive force and the potential difference in the slime mould it is possible to see how the potential rhythm behaves when protoplasm gelates in CO_2 gas. Once protoplasm is brought to the state of a gel, the motive force of the protoplasmic flow can no longer be measured, but the potential rhythm is measurable independently of whether the protoplasm is sol or gel. It was shown that when a gas mixture of $90\% \ CO_2 + 10\%$ air is admitted, the potential rhythm, far from disappearing, is conspicuously amplified during the period in which the whole protoplasm is in the state of a gel (KAMIYA and NAKAJIMA 1955). Such a fact also confirms the above conclusion in terms of bioelectric activity.

Thus we may discard the view postulating physical or chemical changes, which might accompany the displacement of endoplasm, as the cause of reversal. Periodic functioning has its origin in a deeper source.

From this standpoint, the observation made by HOFFMANN-BERLING (1954, 1955) deserves special attention. Among various kinds of interesting experiments on a 'cell model' studied by him, he found the remarkable fact that isolated sperm tails which had been kept in cold 45% glycerol for two weeks have still the capacity to perform their typical oscillatory motion if ATP is added as an energy source. The frequency of oscillation increases with the concentration of ATP. This fact may imply that, for the mechanism of periodic motion, the existence of undenatured actomyosin-like protein, ATP and certain amounts of K^+ and Mg^{++} suffices. So far there has been no report on the dynamic properties of the glycerol-extracted plasmodium to which ATP and certain metal ions have been added, although there are some unpublished data obtained by us.

Recently a new fact was revealed by KISHIMOTO (1958 b) in connection with the electric potential rhythm of *Physarum.* If, by establishing a pressure difference artificially between the two compartments of the double-chamber, the plasmodium is forced to flow for some time with an arbitrary period back and forth—a period which is different from the endogenous rhythm of the slime mould—the modified period is seen to last over several rhythms in the potential change even after the external pressure is removed. Since the period of the DPG and potential rhythm (EPG) are always the same (KAMIYA and ABE 1950), it is reasonable to assume that the rhythm of the protoplasmic streaming must also be modified for a definite period, the modified rhythm duration depending upon the pre-treatment. This is an interesting fact and the problem awaits further study.

2. Rhythmicity of protoplasmic motion in plant cells

The periodic activity of protoplasm discussed so far is concerned only with slime moulds. It is, however, a question, and an important one, whether or not rhythmic nature is a specific attribute of slime mould protoplasm or whether it is essentially universal and, therefore, is found in other protoplasmic systems as well.

In fact, it may be that periodicity of protoplasmic activity is a characteristic common to many other protoplasmic system, but which nevertheless does not manifest itself as a visible phenomenon. Only in special cases like the slime mould, may this periodicity come in evidence externally. The fact that a visible form of movement does not show periodicity cannot offer any cogency to the argument that the basic mechanism is not periodic. If the periodic activity of each minute structural unit did not synchronize, periodicity in the total mass would, as a result of *intraplasmic interference,* tend to be cancelled out statistically.

a) *Bacillaria*

Viewed in this light, the movement of the diatom *Bacillaria paradoxa* is interesting. This diatom is found in colonies consisting of many or many tens of cells which, wonderfully enough, are closely connected with one another by extra-membraneous protoplasm. And the adjacent cells are being shifted parallel to each other through the action of this extra-membraneous living protoplasm so as to change their relative positions. This is a back and forth movement having a definite period. The summation of these changes in the relative positions of each individual organism brings about a remarkable periodic expansion and contraction of the colony as a whole. In such a protoplasmic system we can observe most clearly as a visible phenomenon the periodicity of the movement and the phases of the activity in different loci. A point to be noticed here is that the phase of the shift between the cells may either be the same over the whole protoplasmic system or it may be different locally so that the colony as a whole does not keep a uniform rhythm. *Bacillaria paradoxa* is thus one of the good examples demonstrating a case in which the phase of activity within a continuous protoplasmic system shows local discrepancies.

b) *Euglena*

Another interesting instance is the phenomenon seen in *Euglena.* Some *Euglenae, e. g., E. deses,* exhibit a characteristic change in shape and attendant streaming of the endoplasm, which is generally known as euglenoid movement or *Metabolie* in German. Under normal conditions the movement is irregular and with no indication of rhythm. It is, however, worth noticing that when some *Euglenae* are transferred into acid solutions or subjected to other unfavourable conditions the movement soon begins to show almost perfect periodicity for a considerable time before death takes place (KAMIYA 1939). The rhythm is in this case pathological, but it is noteworthy as an example showing that a movement that was not periodic at all in normal conditions may turn into a periodic motion.

c) *Spirogyra*

It was shown by FREY-WYSSLING (1940) that the streaming cytoplasm in *Spirogyra*-cell is capable of performing rhythmic contraction. When a cell with circulating cytoplasm is treated with 50% methanol, plasmoptysis occurs. The expelled cytoplasm then shows periodic convulsive contractions for more than a quarter of an hour. He observed that this extruded and still mobile cytoplasm exhibits a pronounced double refraction. FREY-WYSSLING (1955) interprets this fact by postulating the existence of contraction waves in the protoplasm which have the appeareance of convulsive contraction rather than of streaming.

d) *Nitella*

As mentioned at the beginning of this chapter, there is no sign of rhythmicity in the case of rotation such as represented by the *Nitella* cell. That rhythmicity is apparently lacking, however, does not mean that no rhythmic process is participating in the formation of that movement. This will become clear when we consider the macroscopic movement of animals. Though the animal body does not show any periodicity in its locomotion when advancing, this movement is formed by repeated activities of special locomotory organs in each body such as cilia, flagella, and many other macroscopic organs. In protoplasmic streaming, we cannot see directly the mechanism causing it, nor is our knowledge about its physical nature any better than mere conjecture. In spite of that, it is almost impossible to understand a regular streaming such as rotation without resorting to the repeated activities of an unknown mechanism whether it may be sol-gel transformation, hydration-dehydration, contraction-relaxation, etc.

In connection with this, it is of great interest that the isolated drops of protoplasm, squeezed out forcefully by YOTSUYANAGI (1953 b) from a cell of *Characeae* into a salt solution within a certain range [0.20–0.25 M $Ca(NO_3)_2$ or $CaCl_2$], showed rhythmic pulsation. Among many types of movement YOTSUYANAGI distinguishes three types of patterns of rhythmic pulsation.

The first type is the simplest form of the rhythmical pulsation of the protoplasmic membrane. A certain part of the surface of a protoplasmic drop advances to the outside in a direction perpendicular to this surface, then contracts itself immediately to regain the original place, and repeats these movements many times successively, often with a very short period (Fig. 36 *a*). This movement takes place without the pulsating part being shifted.

The second type is the following: a part of the membrane makes the same movement as that of the first type, and at the same time this pulsat-

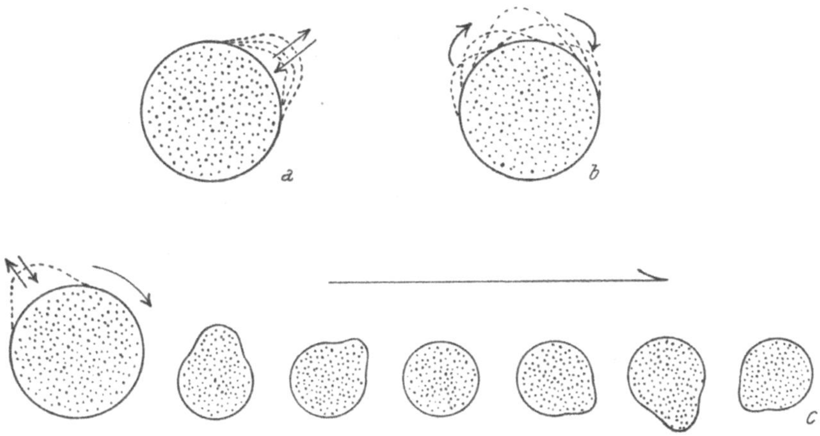

Fig. 36. Pulsation of an isolated drop of protoplasm from a cell of *Characeae*. *a*: Rhythmic contraction at a part of the protoplasmic surface. *b*: The pulsating part is transmitted on the protoplasmic surface within a limited range. *c*: The pulsating part turns around a protoplasmic drop in a kind of undulation. (YOTSUYANAGI 1953 b.)

ing part shows some shifting as an undulation on the surface of the protoplasmic piece. This undulation appears at a part of the membrane, is shifted a little, then disappears at another part, and begins again the same process from the original point (Fig. 36 *b*). The movements of the first and second types seem to be unstable, and after having lasted during some minutes, they are transformed into a movement of the third type.

In the third type, the contractile undulation is propagated along the circumference of a protoplasmic drop, and thus continues to turn around the drop repeating simultaneously its rhythmic contraction in a direction perpendicular to the surface (Fig. 36 *c*). This type of movement, which is more stable than the preceding two, has been observed sometimes for more than 40 minutes in drops of 10 μ to 20 μ in diameter.

We have often met cases when a certain number of contractile waves move along the surface of a protoplasmic drop; consequently the protoplasmic piece is deformed periodically, taking some definite figures alternately.

It is to be noticed that the isolated drops of protoplasm observed by

YOTSUYANAGI consist not only of endoplasm but contain cortical plasm, and it seems to the author that the latter is responsible for the above behaviour. At any rate the behaviour described by YOTSUYANAGI is interesting because it implies that the protoplasm, which, when it is in an intact cell, shows only a regular rotational streaming with a stabilized speed, has the potentiality to be converted so as to show a rhythmic motion. We are now in a position to suppose on the ground of the striking experimental evidence that periodic functioning virtually exists in the protoplasm exhibiting rotation. But it must be concealed in its natural state beneath the microscopic level.

VI. Energetics of Protoplasmic Streaming

Just as in the contraction of muscle, movement of amoeba, cilia etc., protoplasmic streaming is an energy-consuming phenomenon. It is an indispensable condition for the streaming that the chemical energy, that is liberated through metabolism, is changed into mechanical energy. Generally, the energy of a substrate is transferred by way of respiration or fermentation to the phosphate bond of adenosine triphosphate (ATP) which is the energy donator for most of the energy-consuming physiological activities. The physiological meaning of respiration or fermentation lies therefore in no other than the production of ATP. For the time being, however, we shall consider how the protoplasmic streaming is dependent upon respiration and fermentation.

1. Oxygen tension

Much work has been devoted for many years to the problem of the energy source of the protoplasmic streaming. Many authors believe that oxygen is indispensable for the streaming, since in many plant cells the flow stops in a medium which is deprived of oxygen. But sensitivity of the protoplasmic streaming to lack of oxygen is very different according to the type of cells or their age.

It was known a long time ago that the streaming in the staminal hair of *Tradescantia* stops in an atmosphere of pure hydrogen within from 15 minutes to 2 or 3 hours according to the temperature and also to the age and condition of the cell (cf. EWART 1903). According to KELSO and TURNER (1955), the rate of protoplasmic streaming in the staminal hair of *Tradescantia virginica* is not affected by change of oxygen concentration within a range of 1.5–100 per cent. In their experiments the rate of flow was first determined when the staminal hair-cells were enclosed in a hanging drop over air. Then the new steady rate was measured in a different gas mixture of oxygen and nitrogen. Although 1.5% oxygen (the critical value) has no effect, the rate rapidly decreases to a new steady level, when concentration of oxygen is further reduced. Protoplasmic streaming is brought to a complete standstill within 15 minutes in hydrogen gas containing 0.3 per cent oxygen. Recovery of flow takes place, however, without delay by the re-administration of oxygen.

The effect of low oxygen tension has been especially well studied in the coleoptil cells of *Avena*. The streaming in this material requires oxygen to such a great extent that it is necessary for longer observation to use aerated water or to substitute a new medium frequently. If a young coleoptile, 4–5 cm. long, is split lengthwise and mounted in pure water under a coverslip, the rate of flow begins to decrease within a few minutes. BOTTELIER (1935) attributed this fall in the rate to the deficiency of oxygen. The streaming rate is restored to the initial level when a stream of fresh water is passed under the coverslip.

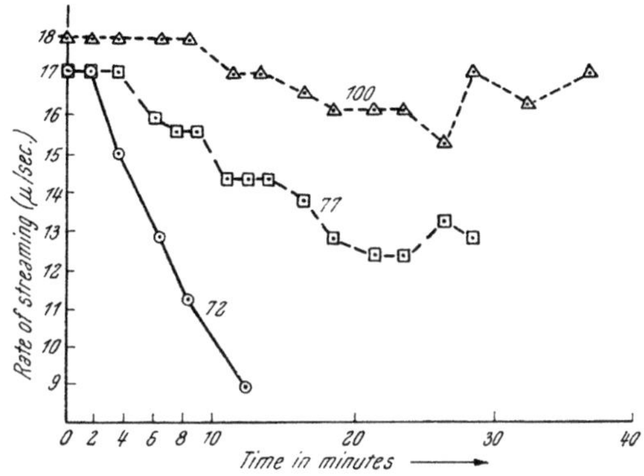

Fig. 37. Streaming rate in epidermal cells of *Avena* coleoptile placed in limited quantities of water. Coleoptiles 72, 77, and 100 hours old respectively, each in 0.4 ml. aerated water, not renewed.
(SWEENEY and THIMANN 1938.)

It has been repeatedly demonstrated for *Avena* coleoptile that the younger cell is more sensitive to oxygen deficiency than the older one (BOTTELIER 1934, 1935; EYMERS and BOTTELIER 1937; SWEENEY and THIMANN 1938; DUBUY and OLSON 1940 (cf. Fig. 37). BOTTELIER showed that for young coleoptiles (grown for 96 hr. at 26° C.) the streaming rate increases if the surrounding water charged with air is replaced with that saturated with oxygen. In case the critical tension of oxygen for the cell is low, a rapid decrease in streaming in unaerated or unchanged water drops is seen only when comparatively large (respiring) samples of tissue are used for the experiment.

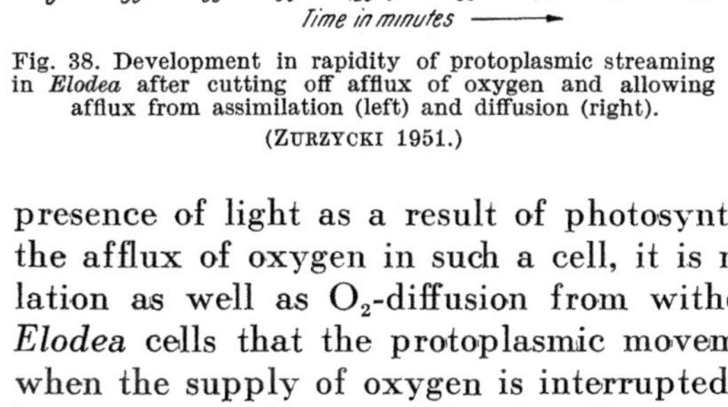

Fig. 38. Development in rapidity of protoplasmic streaming in *Elodea* after cutting off afflux of oxygen and allowing afflux from assimilation (left) and diffusion (right).
(ZURZYCKI 1951.)

In a green cell containing chloroplasts, oxygen is supplied through diffusion from the external medium or by intracellular production in the presence of light as a result of photosynthesis. Therefore, for cutting off the afflux of oxygen in such a cell, it is necessary to interrupt its assimilation as well as O_2-diffusion from without. ZURZYCKI (1951) observed in *Elodea* cells that the protoplasmic movement stops within 30–60 minutes when the supply of oxygen is interrupted. The streaming is resumed and kept at a steady rate either by illuminating the leaf when the afflux of oxygen from without is cut off, or by supplying oxygen from the medium while assimilation is interrupted. We can therefore conclude that the oxygen required for the streaming can be derived both from assimilation and diffusion from the outside although in the latter case the rate is not so great as in the former (Fig. 38) (cf. also p. 92).

Sensitivity to lack of oxygen is very variable among different types of cells. Among early works, there are the careful experiments of EWART (1903) who reported that protoplasmic streaming continues, though slowly, in healthy plants of *Chara foetida, Nitella translucens, N. flexilis,* and to a less extent in *Nitella syncarpa, Chara gracilis, C. flexilis,* and *C. stelligera,* for from one to several weeks at temperatures from 15° C. to 18° C. in darkness in the "entire absence of oxygen."

The protoplasmic streaming in myxomycete plasmodium shows under low oxygen tension quite a different behaviour from many of the plant

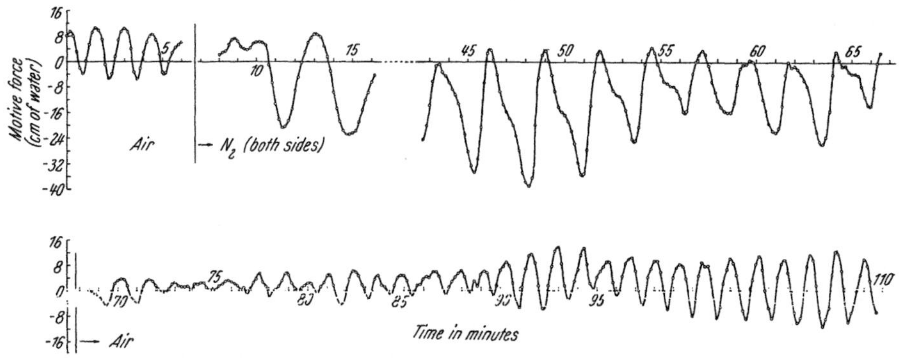

Fig. 39. Generation of the motive force in the plasmodium in and after anaerobiosis. (KAMIYA 1953.)

cells having cell walls. Among the related literature that appeared in the last two decades there are the works of KITCHING and PIRENNE (1940), PRICE and ALLEN (1942), ALLEN and PRICE (1950), LOEWY (1950), OHTA (1952 b), KAMIYA (1953), KAMIYA, NAKAJIMA and ABE (1957), KAMIYA, ABE and NAKAJIMA (1958), etc. All of these authors studied the same material, the plasmodium of *Physarum polycephalum.* KITCHING and PIRENNE reported that streaming ceases under complete lack of oxygen but only a small partial pressure (1–3 mm. Hg column) of oxygen is sufficient for maintaining or restarting the streaming. According to ALLEN and PRICE (1950), the streaming is normal until the oxygen pressure reaches 2.4% of the atmospheric pressure; the rate of flow diminishes slightly under 1% oxygen; under 0.3% oxygen, streaming still continues while growth can no longer occur. Even under complete lack of oxygen, streaming lasts for several hours. On the other hand, LOEWY (1950) observed that the plasmodium of *Physarum polycephalum* disintegrates in pure nitrogen in 20–80 minutes, but a small quantity of CO_2 removes the toxic effect. Namely, in the mixed gas of 95% N_2 + 5% CO_2, the streaming continues in spite of complete lack of oxygen, for more than 24 hours without showing any ill effect. LOEWY, therefore, came to the conclusion that the lack of carbon dioxide, and not the absence of oxygen itself, is responsible for the cessation of flow under pure nitrogen. This is a fact to be noted, but here we are again confronted with the historical problem as to to what extent the cessation of flow is due to the fading out of the motive force back of the streaming and to what extent it is due to the increase in viscosity.

In order to see how the low oxygen tension affects the motive force of the streaming OHTA (1952 b), KAMIYA (1953), and KAMIYA, NAKAJIMA and ABE (1957) mapped dynamoplasmograms of *Physarum polycephalum* under lack of oxygen using the double-chamber method. As nitrogen gas is not perfused but kept enclosed in the two compartments during all of these experiments, a small amount of CO_2 produced by the organism as a result of fermentation (OHTA 1954) may be contained in the two compartments. A possible toxic effect due to the complete absence of CO_2 may thus be excluded. The response of motive force production to the lack of oxygen may be different to some extent according to the condition of the plasmodium especially with regard to the state of nutrition, but in no case does the motive force tend to decrease under low oxygen tension. The results obtained by KAMIYA (1953), KAMIYA, NAKAJIMA and ABE (1957) showed a striking increase in the motive force in nitrogen gas (Fig. 39).

The recent work of KAMIYA, ABE and NAKAJIMA (1958), in which both the motive force of the streaming and respiration were measured simultaneously by means of a device described in the previous pages (49 ff.), revealed that in 2.5% oxygen the amplitude of dynamoplasmogram increases about twice as much as that obtained under the normal state (20% O_2) while the oxygen uptake decreases to half of the normal level. In the gas containing more than 5% oxygen, there is no effect any more both in respect to respiration and motive force generation. Sensitivity of respiration and that of motive force generation to the lack of oxygen are about the same, but it is remarkable enough that the motive force increases rather than decreases under reduced oxygen tension, while respiration suffers reduction under the same condition.

2. Metabolic inhibitors

a) Cyanide

The effect of cyanide and carbon monoxide on the protoplasmic streaming is very similar to that of oxygen deficiency. SWEENEY and THIMANN (1938) have shown that cyanide, at 1×10^{-4} and 5×10^{-4} molal, reduces the streaming rate. The effect of cyanide is quick, but draining off the cyanide does not immediately restore the initial rate. It takes 5–10 minutes before the normal rate is resumed.

According to DUBUY and OLSON (1940), 3×10^{-4} M KCN reduces the respiration rate of *Avena* coleoptile to about 50% while concentrations up to 1×10^{-2} M KCN reduce it to the same degree. As for the protoplasmic streaming, it is also inhibited by KCN, but, as in the case of respiration by only about 50% even up to the concentration of 1×10^{-2} M (OLSON and DUBUY 1940) (Fig. 40). The inhibition of the streaming caused by KCN is reversible provided the material is not exposed to a high concentration for a long time. These authors suppose that two respiratory systems are effective for producing protoplasmic streaming: the one being cyanide-sensitive, the other cyanide-indifferent. At any rate, it is noted that under the influence of KCN the reduced rate of respiration and that of flow go parallel to each other in *Avena*.

DOI (1950) found similar, but somewhat different results in protoplasmic streaming in root hairs of *Oryza, Triticum, Hordeum,* which are very sensitive to dilute KCN solutions. The rate of flow in the rice plant diminishes to nil in 10^{-4} M KCN (pH 6.6) within 10 minutes. Even in 10^{-5} M KCN (pH 6.6) the retarding effect is obvious, though the streaming does not stop completely. In a strain of wheat, the streaming stops in 20–30 minutes in 10^{-3} M KCN (pH 7.2), and it becomes quiet in 50–60 minutes in 5×10^{-4} M. Thus it is clear that a cyanide-sensitive respiration system is responsible for the streaming. Cyanide has an effect similar to that of low oxygen tension on the streaming in these root hairs.

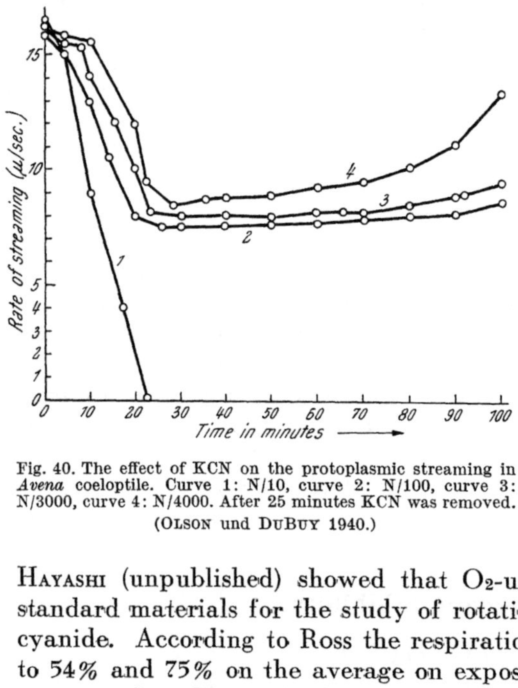

Fig. 40. The effect of KCN on the protoplasmic streaming in *Avena* coeloptile. Curve 1: N/10, curve 2: N/100, curve 3: N/3000, curve 4: N/4000. After 25 minutes KCN was removed. (OLSON und DUBUY 1940.)

As for cells having chloroplasts, GIMESI and POZSAR (1955) studied the effect of KCN on the rate of flow in *Elodea* leaf. The rate of flow is diminished in KCN solutions of 10^{-4} M, 10^{-3} M and 10^{-2} M to 89%, 76% and 68% of the control respectively.

ROSS (1934, 1938) and HAYASHI (unpublished) showed that O_2-uptake of *Nitella or Chara* cells, standard materials for the study of rotation, are inhibited considerably by cyanide. According to ROSS the respiration of *Nitella clavata* is decreased to 54% and 75% on the average on exposure to 10^{-3} M and 10^{-4} M NaCN respectively. HAYASHI's data on *Chara Braunii* show a slightly stronger inhibition of O_2-uptake with the same concentration of KCN (pH 7.3). The rate and the motive force of flow in *Chara Braunii*, however, are not affected by KCN, at least up to the concentration of 10^{-3} M (HAYASHI 1953).

Protoplasmic streaming in the myxomycete plasmodium behaves rather differently under the influence of cyanide from that in many plant cells. It is known that the respiration of the plasmodium of *Physarum polycephalum* is very sensitive to cyanide (ALLEN and PRICE 1950, OHTA 1954). An appreciable inhibition was observed at 5×10^{-4} M. There still remains, however, a residual respiration of about 20% even at 10^{-2} M KCN. On the other hand, it has been reported by OHTA (1952 b) and by KAMIYA, NAKAJIMA and ABE (1957) that the motive force of the streaming increases rather than decreases under the effect of cyanide as was also the case under low oxygen tension (Fig. 41). By their method of simultaneous measurement of the motive force and respiration in the plasmodium, Kamiya, Abe and Nakajima (1957 b) found that residual O_2 uptake becomes 60% of the control under the influence of 2×10^{-3} M KCN while the motive force augmented to such a great extent that its amplitude became about twice as large as that of the control.

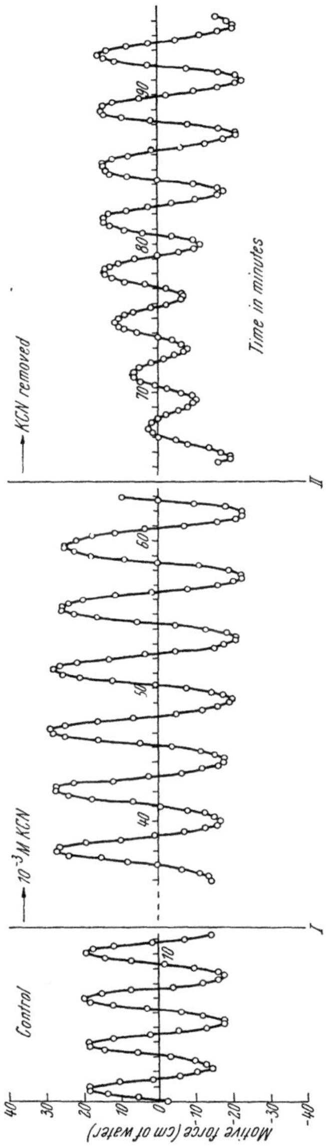

Fig. 41. Generation of the motive force in the plasmodium under the effect of 10^{-3} M KCN. Both sides of the plasmodium in the double-chamber were treated evenly during the period between I and II. (Kamiya, Nakajima and Abe 1957.)

b) Carbon monoxide

It is well known that cytochrome oxidase system is also inhibited by CO and that this inhibition is removed upon illumination with white light. The effect of CO on motive force generation in *Physarum* plasmodium, which has been studied by Ohta (1952 b) and Abe, Nakajima and Kamiya (unpublished), is basically the same as that of cyanide, *i. e.*, CO augments the force back of the protoplasmic streaming. The experiments were conducted keeping the double-chamber in darkness except a limited spot on the connecting strand which is allowed to be illuminated for observation. The latter authors observed a tendency in the augmented motive force to resume the normal magnitude upon illumination and Allen and Price (1950) and Ohta (1954) observed a "nearly complete" or "slight" recovery in respiration upon illumination.

c) 2, 4-Dinitrophenol (DNP)

The effect of 2, 4-dinitrophenol (DNP) on the protoplasmic streaming had been studied in *Avena* (Sweeney and Thimann 1938, Olson and DuBuy 1940) before it was known that this reagent uncouples phosphorylation from oxidation. According to Sweeney and Thimann, the streaming rate falls rapidly in the concentration of 100 mg./l. (5.4×10^{-4} M) and less rapidly at 10 mg./l. The decreased rate begins to return to normal as soon as the substance is removed. Although DNP causes no observable decrease in streaming rate any longer in the concentration 1 mg./l. it decreases the rate if auxin is added at the same time. Sweeney and Thimann say that all auxins at any concentration decrease the streaming rate in the presence of DNP. The results obtained by OLSON and DUBUY (1940) about the DNP-effect on the streaming is more or less similar to those obtained by SWEENEY and THIMANN (1938). As shown in Fig. 42, DNP at the concentration of 100 mg./l. causes a rapid fall in streaming rate reaching complete stop in about 20 minutes, after which no recovery occurs. In the concentration of 50 mg./l., the rate decreases to about 50% and remains unchanged for a long time even after the solution is withdrawn. With a concentration of 10 mg./l., which decreases the rate to about 70%, recovery occurs in 10 minutes or so with the removal of the reagent.

DNP was also applied by PFEIFFER (1953) to the staminal hair of *Tradescantia* in order to suppress the flow reversibly. The concentration of 35 mg./l. (1.8×10^{-4} M) reduces the rate of streaming from 10–12 μ/sec. to 5.5 μ/sec. SIBAOKA and ODA (1956) used DNP for their experiments on action potential and found that 5 and 9×10^{-3} M DNP inhibit the streaming in *Chara* internodes, but was no longer effective at 10^{-3} M.

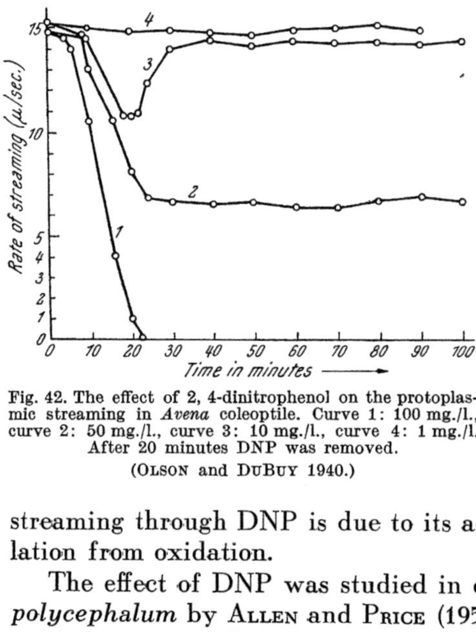

Fig. 42. The effect of 2, 4-dinitrophenol on the protoplasmic streaming in *Avena* coleoptile. Curve 1: 100 mg./l., curve 2: 50 mg./l., curve 3: 10 mg./l., curve 4: 1 mg./l. After 20 minutes DNP was removed. (OLSON and DUBUY 1940.)

We thus understand that DNP has a marked inhibitory effect on the streaming and this effect is reversible if the concentration of the solution is not high or if the duration of action is not too long. Perhaps the inhibition of the streaming through DNP is due to its action as an uncoupler of phosphorylation from oxidation.

The effect of DNP was studied in detail in the slime mould, *Physarum polycephalum* by ALLEN and PRICE (1950), KAMIYA, NAKAJIMA and ABE (1957), etc. ALLEN and PRICE (1950) showed that respiration is stimulated by 2×10^{-4} M and 1.25×10^{-3} M DNP, so as to increase by 20–50%, yet protoplasmic streaming stops within a few minutes after 2×10^{-4} M DNP is admitted. It has been further shown by KAMIYA, NAKAJIMA and ABE (1957) that the inhibition of the streaming through DNP is not due to the gelation of the protoplasm, but to the abolition of the motive force production. Fig. 43 shows how the motive force is changed under the effect of 5×10^{-4} M (pH 7.0). The two parts of the plasmodium in the double-chamber are treated evenly by DNP in this case. The waves in the first 35 minutes represent the normal activity of the plasmodium. During the period of 42 minutes between the two vertical lines I and II, the plasmodium was bathed in 5×10^{-4} M DNP. As illustrated in the figure, during the first 10 minutes or so after the material was subjected to the reagent, the motive force was still produced to an even greater degree than in the control, but the force eventually wore off. Decrease in the motive force is accompanied with decrease in viscosity, a fact which can be ascertained, as mentioned before, by examining the rate of protoplasmic flow induced

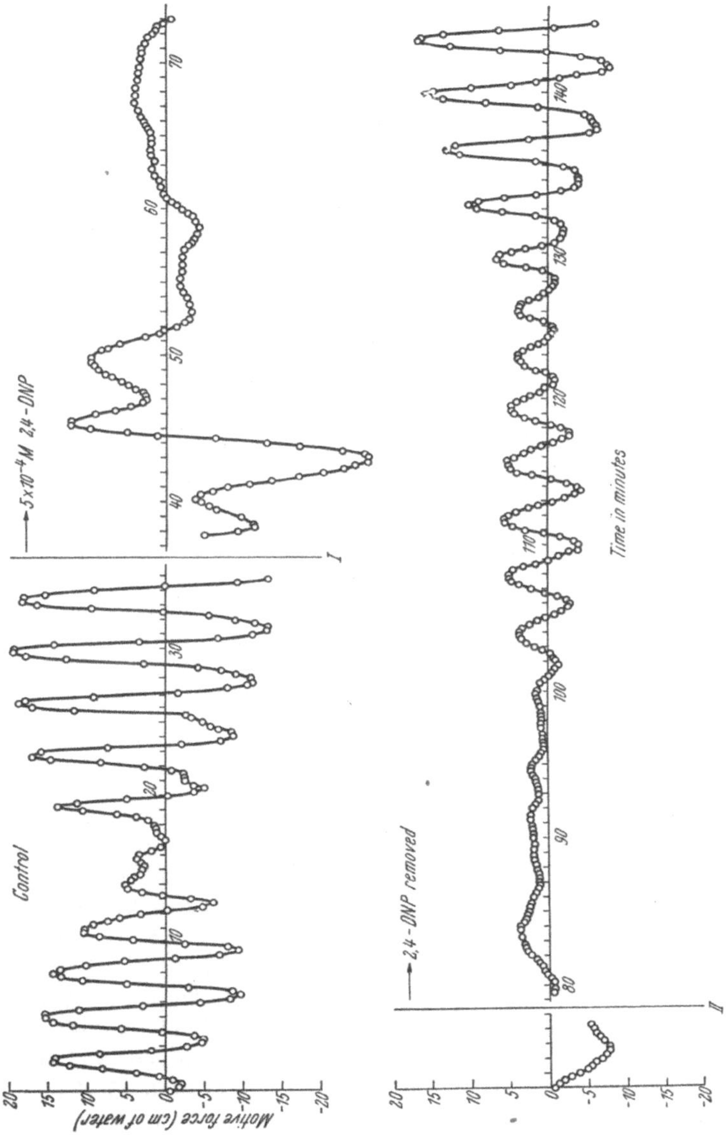

artificially in the connecting strand between the two compartments. After the removal of the reagent, the production of the force was not resumed promptly, but the gradual and complete recovery of the rhythmic motive force is clearly seen in the graph. Therefore the pronounced effect of DNP on the streaming is perfectly reversible.

Perhaps one might be inclined to think from the above behaviour of the slime mould in DNP that oxidative phosphorylation is essential to the protoplasmic flow also in this organism. It is, however, to be noted that the motive force decreases also in the anaerobic state through the effect of DNP in the same concentration ($5 \cdot \times 10^{-4}$ M) as is used in the aerobic state, a fact which is not to be overlooked inasmuch as DNP is known as an uncoupler of the oxidative phosphorylation. Probably, as far as the slime mould is concerned, the suppression of the motive force under DNP is not brought about through the uncoupling of oxidative phosphorylation. This conclusion is well in conformity with the results of the experiment performed by TAKEUCHI and HATANO (unpublished) who showed that the level of ATP is lowered conspicuously under the influence of DNP no matter whether the organism is in aerobic or anaerobic state. In Fig. 43 it is noticed that the amplitude tended at first to increase before it decreased. Although such an incipient stimulation does not always appear in different samples, it is possible that the transient increase of the force at the beginning is caused by stimulation of ATPase activity of plasmodial myosin B through DNP (NAKAJIMA, unpublished).

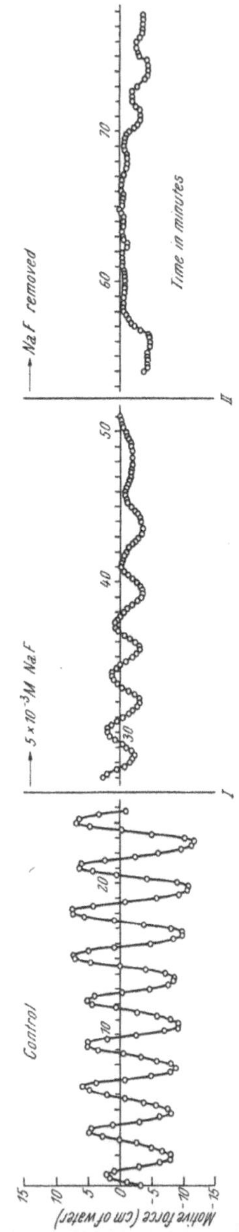

Fig. 44. The effect of 5×10^{-3} M NaF on the production of the motive force. The plasmodium in the double-chamber was treated evenly during the period between I and II. (KAMIYA, NAKAJIMA and ABE 1957.)

d) Monoiodoacetate (MIA) and fluoride

There are but few data available on the effect of fermentation poisons, such as monoiodoacetate, fluoride, etc. on the protoplasmic streaming in plant cells. The former inhibits dehydrogenation from triosephosphate and the latter dehydration of 2-phosphoglyceric acid.

SWEENEY and THIMANN (1942) reported that iodoacetate completely inhibits the acceleration of streaming by auxin in the concentration which inhibits growth (5×10^{-5} M). But the basal streaming rate without auxin is not affected by higher concentrations up to 2×10^{-4} M.

GIMESI and POZSÁR (1955) showed that the rate of flow in an *Elodea* cell is suppressed to 50% of the control by sodium fluoride at a concentration of 1×10^{-3} M, whereas stimulation is seen to be nearly 140% of the basal streaming rate at lower concentrations around 1×10^{-4}—2×10^{-4} M. Whether or not this acceleration is due to the increase in the motive force or to the decrease in viscosity is unknown. Probably the latter may be the case.

The effect of both MIA and sodium fluoride on the motive force generation has been studied in the slime mould. In both cases the production of the motive force is markedly inhibited at a concentration at which respiration is not appreciably affected. The phenomenological difference in the effect of monoiodoacetic acid and that of NaF is that the viscosity of protoplasm increases in the former case while it decreases in the latter case.

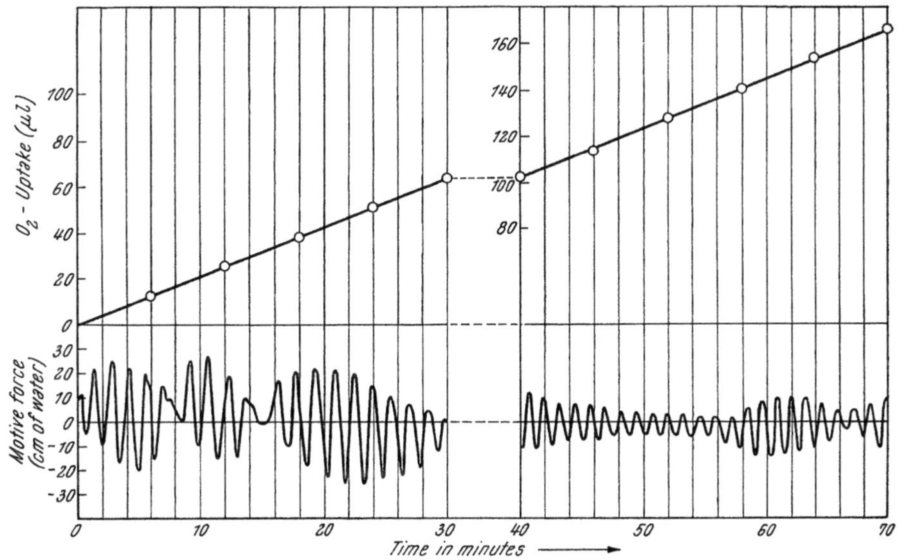

Fig. 45. Simultaneous measurement of the motive force of the protoplasmic flow and oxygen-uptake of the plasmodium before (left half) and after (right half) treatment with 4×10^{-4} M MIA. (ABE, NAKAJIMA and KAMIYA 1958.)

Fig. 44 shows the effect of 5×10^{-3} M NaF. Upon replacement of the external buffer solution with that containing NaF, the motive force decreases immediately to ¼ the normal value although the period of the wave suffers well-nigh no change. It is known that respiration is not much altered by sodium fluoride in the concentration used (OHTA 1954). From Fig. 44 it is seen that the motive force did not recover quickly upon removal of NaF.

The motive force of the streaming in the plasmodium also undergoes a remarkable suppression through MIA. Fig. 45 shows an example of simultaneous measurements of the motive force and respiration of a plasmodium when it was treated as a whole with 4×10^{-4} M MIA (pH 7.0). We see that the motive force decreases conspicuously as soon as the reagent is admitted without inhibition of respiration ($Q_{O_2} = 8.0$).

It is to be noticed that the actions of fluoride and MIA are followed by the suppression of the motive force without delay. It is interesting to refer in this connection to the recent experiment of TAKEUCHI and HATANO (unpublished), who showed that there is no phosphocreatine or phosphoarginine detectable in the slime mould. Moreover, according to NAKAJIMA (1957) MIA does not influence the ATPase activity of the plasmodial myosin B (cf. p. 160). Therefore, the rapid decrease of the motive force under the effect of MIA or fluoride may be accounted for by the absence of the immediate reserve cycle, such as is found in muscles of higher and lower animals in the form of phosphocreatine or phosphoarginine.

Reviewing all the data so far obtained by KAMIYA *et al.* on the effect of metabolic inhibitors on the motive force production in the slime mould, we realize that the conditions which suppress respiration increase without exception the force back of the streaming. If the slime mould were free to flow, the rate is likely to increase under these conditions, a fact which, however, was not noticed by previous authors. Such a fact has never been observed in plant cells having a cell wall and indicates a rather peculiar nature of the slime mould in respect to the energetics of the streaming. On the other hand, conditions which block the glycolytic process (EMBDEN-MEYERHOF-PARNAS passway) invariably diminish the motive force when respiration is not affected. Though the problem as to what process is attacked in a reversible fashion by DNP in the slime mould still awaits further study, the data so far obtained by KAMIYA *et al.* indicate that the energy required for the streaming is supplied by a glycolytic process. Viewed from this standpoint, the increase in the motive force in the anaerobic state or under the influence of KCN or CO can be understood as reflecting increase in glycolysis due to the PASTEUR effect, which has been observed in this material by OHTA (1954).

All in all, from young plant cells up to the myxomycete plasmodium we realize that there are cases in which oxygen-dependence of protoplasmic flow can be widely different.

3. The role of ATP in the protoplasmic flow

a) ATP as the energy source for the motive force production in the myxomycete plasmodium

In view of the fact that the energy required for most of the physiological phenomena is supplied through energy-rich phosphate bond represented by adenosine triphosphate (ATP), it is important to know what happens to motive force generation when ATP is admitted to the plasmodium either from the outside or by microinjection. Experiments along this line have been done by KAMIYA and his collaborators.

In Fig. 46 is shown an example in which only one half of a plasmodium was treated with 2×10^{-3} M ATP. The two compartments *A* and *B* were both filled with 0.02 M phosphate buffer (pH 7.0) at first; at the vertical line *I* marked at 20 minutes on the time scale, 2×10^{-3} M ATP dissolved in the same buffer solution was substituted for the solution in compartment *A*. The change in the wave pattern after ATP was admitted is very characteristic. Though there appeared no immediate change in amplitude, period and polarity of the wave, the wave train began to show a pronoun-

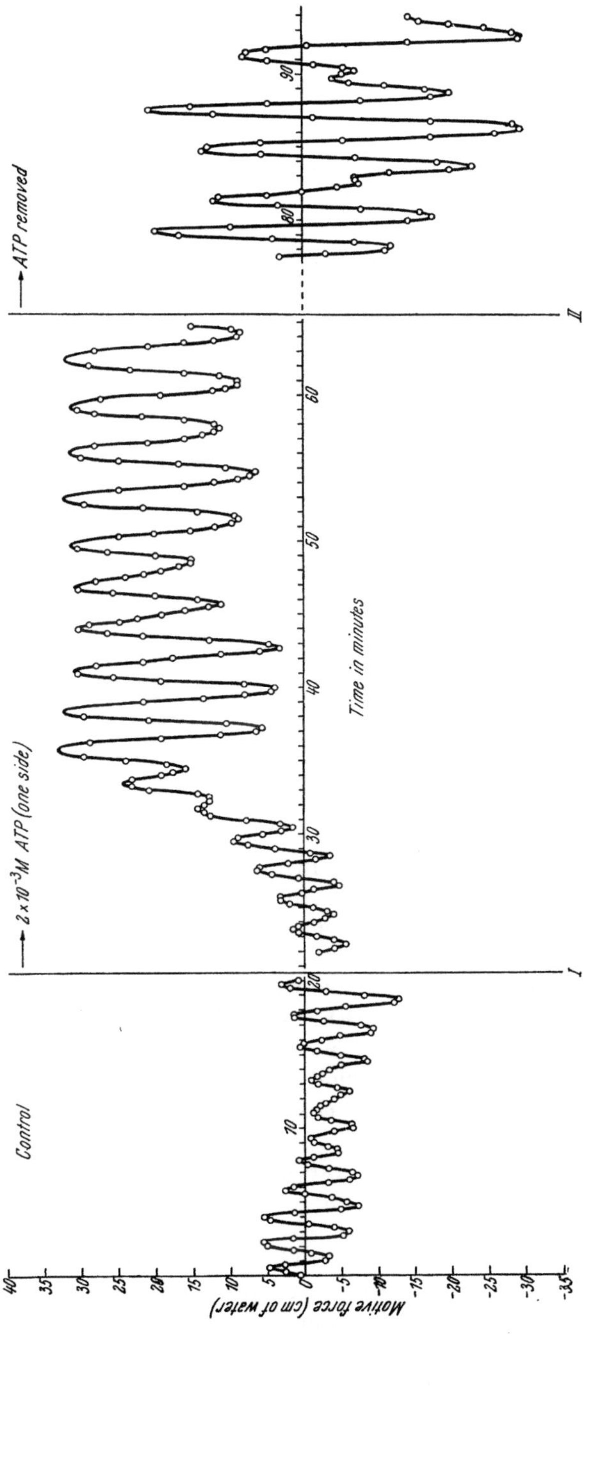

Fig. 46. Dynamoplasmogram under the influence of 2×10^{-3} M ATP admitted to one half of the plasmodium (*a* in Fig. 22, p. 41) at I. ATP was removed at the time designated as II.
(KAMIYA, NAKAJIMA and ABE 1957.)

ced tendency to take an upward course 6 to 7 minutes after ATP was admitted to the protoplasmic blob in the left compartment (*A* in Fig. 22). As the curve represents the pressure in *B* just sufficient to keep the endoplasm in the connecting strand at a standstill, the fact that the wave train goes upward and waves no longer cross the base-line is an indication that the inner pressure in the blob treated with ATP is increased to such a great extent that the force is exerted always in one direction without reversal, *i. e.*, from *a* to *b*. If there had been no balance-pressure in action, the endoplasm would have started to stream from the ATP side to the side of plain buffer solution. 15 minutes after ATP was admitted to the left half of the plasmodium, the motive force repeats its periodic activity without taking further upward course with an amplitude that is conspicuously augmented. By replacing the ATP solution in compartment *A* with the plain buffer solution, the waves as a whole took a downward course again, but their amplitude did not show a tendency to decrease even 25 minutes after ATP was removed.

As for the viscosity of protoplasm, which can be estimated by the sensitivity of the protoplasmic flow artificially caused by pressure difference, it was noticed in this experiment that no immediate change followed after ATP was admitted. Only several minutes later, that is, accompanying the increase in the motive force, viscosity also increased. The increase in viscosity persisted also after ATP was removed.

With respect to the adenylic acid (AMP) of both the muscle and yeast, we could not observe any effect whatsoever on the motive force comparable to that produced by ATP. There are furthermore no discernible changes in viscosity of protoplasm under the influence of AMP. Thus we may assume that the characteristic behaviour of the plasmodium and especially the augmentation of the motive force brought about by ATP are due to the specific action of ATP.

ATP can be administered to the inside of the plasmodium directly by means of the technique of microinjection. It was reported by Ts'o *et al.* (1956 a) that microinjection of low concentration of neutral ATP solutions into plasmodia partially liquefies the plasmagel and increases the rate of streaming. They noticed morphological changes of the plasmodium such as formation of pseudopodium-like protuberances on the periphery, when ATP is introduced in it. In many respects the effects of ATP observed by them in myxomycete plasmodia are in line with those reported by Kriszat (1949, 1951, 1954), and Goldacre and Lorch (1950) or Goldacre (1952) for amoeba.

An effective method for investigating the action of ATP on the protoplasmic streaming is to combine the technique of microinjection and the double-chamber method. If 1 μl. of 10^{-3} M ATP is injected (final concentration: about 1×10^{-4} M) together with K^+ and Mg^{++} in one of the two blobs (ca. 10 mg. each) of the dumbbell-shaped plasmodium placed in a double-chamber designed for that purpose, and if we map the dynamoplasmogram in this case, we get a pattern resembling Fig. 46. The only difference between the two cases is that, when ATP is administered from the outside, it takes 10 or more minutes before the amplitude is increased, while when ATP is injected into the plasmodium, the effect manifests itself without delay. It must be understood that it takes some time for ATP to penetrate the protoplasmic membrane into the interior of the plasmodium in sufficient amount to affect the motive force generation. When ATP is added by microinjection so that its normal level of protoplasm is heightened by *ca.* 10^{-4} M the augmentation of the motive force is maximal. The motive force becomes smaller no matter whether ATP concentration in the protoplasm becomes higher or lower (TAKATA 1957). It is thus reasonable to conjecture that the ATP level of the ground plasm is normally suboptimal rather than optimal or superoptimal in respect to the streaming.

The effect of ATP on the protoplasmic movement of the slime mould has been studied with an entirely different method by KURODA (1958). KURODA cut the slime mould into minute masses as small as 30 μ in diameter and investigated their movement. They perform a complicated movement that is apparently devoid of any order. In order to express this complicated movement quantitatively, a photographic picture was taken of one protoplasmic mass every 15 seconds so as to get its two-dimensional contour. As this protoplasmic mass is changing continuously through a kind of amoeboid movement, the contour that one gets after 15 seconds is somewhat different from the previous one. When these two contours are placed one on top of the other, one can know clearly where the protoplasmic mass advanced and where it receded in 15 seconds. If the sum of the areas covered by these moving contours is obtained and its percentage against the total area is designated as "activity" of the protoplasmic motion, it reaches 3–10%, usually around 5% in 15 seconds in the normal state. The value thus obtained does not of course represent exactly the quantity which stands in direct proportion to the motive force of the streaming, but it is an objective measure of the dynamic activity of a complicated movement. This method also has the merit of increasing the proportion of the outer surface against the volume and renders it easy for the drugs that are applied from the outside to penetrate within the protoplasm.

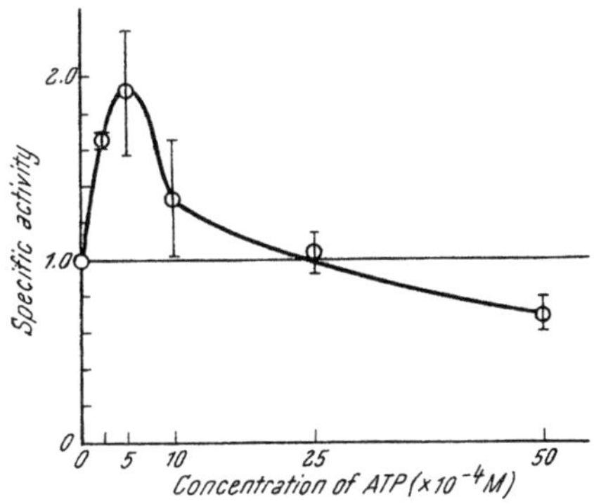

Fig. 47. Relation between the concentration of ATP in the external medium and the motility of small fragments of the plasmodium *(Physarum polycephalum)*. Abscissas: ATP concentration in mol. per liter. Ordinates: Activity of the protoplasmic movement under the influence of ATP in comparison with that in the milieu containing no ATP, bar indicating the mean error. (KURODA 1958.)

Now when ATP is applied from without, is effect differs according to the concentration: at 5×10^{-4} M the movement gets more active after its application and the activity of the motion becomes about 1.9 times as great as before ATP is applied (Fig. 47); at 2.5×10^{-3} M, the viscosity gets higher so that almost no change in activity is observed; at 5×10^{-3} M the viscosity becomes extremely high and the activity gets even lower than before ATP is applied. With AMP, on the other hand, no change in activity of the motion such as was caused by ATP has been observed at any concentration whatsoever.

The results obtained by KURODA agree well with those obtained by KAMIYA, NAKAJIMA and ABE (1957) with the double-chamber method, a fact that shows that the activation of motility through ATP is clearly related with the high energy-phosphate bond.

In this connection, the recent experiments of TAKEUCHI and HATANO (1955, 1956) are of special interest. They showed the existence of ATP in the normal plasmodium of *Physarum polycephalum* at the concentration of 0.4×10^{-3} M. It was also impressively demonstrated by them that ATP obtained from the slime mould contracts the glycerinated muscle (*psoas*) of a rabbit as conspicuously as muscle ATP does.

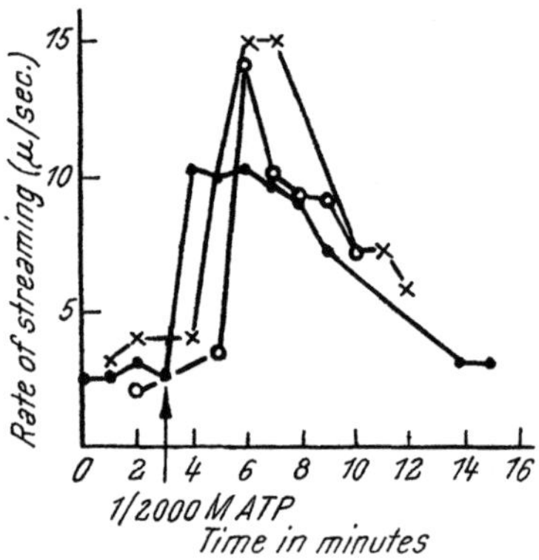

Fig. 48. The effect of ATP (5×10^{-4} M) on protoplasmic streaming in *Acetabularia*. (TAKATA 1958.)

An observation in confirmation of the view that ATP is the direct energy source of the protoplasmic motion was also made in *Amoeba* by GOLDACRE and LORCH (1950) who noticed that the flow of endoplasm in *Amoeba discoides* occurred in the distal direction from the locus where 1–3% ATP was injected with a micropipette. The results obtained on the myxomycete plasmodium further demonstrated this fact on a quantitative basis.

b) Effect of ATP on the protoplasmic streaming in plant cells

As for the effect of ATP on the protoplasmic streaming in plant cells very little is yet known. TAKATA (1958) confirmed, however, that the streaming in the *Acetabularia* becomes, after 5×10^{-4} M ATP (dissolved in sea water) is applied from the outside, four times as active as before it was added. Such an active state, *i. e.*, the state of increased rate, lasts but 10 minutes or so before it returns to the initial rate (Fig. 48). We do not know why the effect is transient. If the concentration of ATP is raised to 2×10^{-3} M, the streaming rate first increases for a brief period of time but it stops eventually, usually 4–5 minutes after the reagent is administered. This reaction is neither brought about by adenylic acid nor by pyrophosphate; hence, it is considered again to be ATP-specific.

4. Availability of ATP

Though so far we are not in a position to claim definitely that the energy required for all kinds of protoplasmic streaming is supplied only by way of ATP, it is reasonable to assume in view of the experimental facts mentioned in the foregoing that this is actually the case so far as the slime mould is concerned. Experiments on the slime mould, however, suggest that ATP produced by respiration with high efficiency is not readily available for the mechanical work of producing the motive force while ATP synthesized by glycolytic process (EMBDEN-MEYERHOF-PARNAS passway) with low efficiency is usable for it.

We may regard both respiration and fermentation as mere means of accumulating energy in the form of the phosphate bond of ATP so that the chemical energy of substrates becomes available through it for various performances of the cell.

Chemically speaking, there is no difference between ATP obtained with high efficiency through respiration and that obtained through the glycolytic process, but it is physiologically significant when considering the availability of ATP as an energy source of the protoplasmic movement that ATP is generated at different parts in the cytoplasm according to aerobic and anerobic metabolism. It is at present a well-recognized fact that the seat of respiration is confined to mitochondria where most of the important respiratory enzymes, *e. g.,* succinic dehydrogenase, cytochrome oxidase, etc. are localized. In the case of plasmodia of *Physarum polycephalum,* too, it is demonstrated that succinoxidase activity is concentrated at the mitochondria (HOLTER and POLLOCK 1952, JOHNSON and MOOS 1956). Glycolysis, on the other hand, takes place in the "soluble part" of the cytoplasm, that is, hyaloplasm or ground plasm. Therefore, it is clear that ATP is produced in mitochondria when the phosphorylation is coupled with aerobic metabolism. On the other hand, when ATP is resynthesized in the glycolytic process, *i. e.,* EMBDEN-MEYERHOF-PARNAS passway, it is produced in the ground plasm.

It is important in this connection to note that the mechanochemical system (p. 160), which is responsible for motive force generation, resides not in cytoplasmic particles like mitochondria but in the ground plasm itself. This means that its availability for protoplasmic motion may not be the same on account of a topographical difference, although ATP in itself is chemically identical. The series of experiments that have been conducted in the slime mould show that only the relatively free ATP that can be obtained through the glycolytic process is utilized as the energy source of the movement.

When we watch the way in which cells like leucocytes and some amoebae move by extending thin pseudopods, we see that the mitochondria do not move at all toward the pseudopodia, but that they are left at the more basal part of the pseudopodia and that only that part of the cytoplasm that is apparently without any visible structure stretches out in length. This fact seems to point out that motion and the mitochondria are not very deeply related to each other. Contrary to this view, there are also cases known in which a close correlation is found between the location of mitochondria and the form or movement of the cell such as is the case in the chick fibroblasts (BIERLING 1955, LETTRÉ and SCHLEICH 1955).

It is an established fact as mentioned already that in the young cells of the higher plants the protoplasmic streaming is strictly dependent on respiration, and it has been observed that even amoebae come to a standstill sooner or later under anaerobic condition. Among the causes accounting for that phenomenon, there may be cases when the products of fermentation induce secondary changes in the physicochemical properties of protoplasm (hydration, viscosity, etc.), under which the movement becomes slack or even stops. It goes without saying, however, that there may be not a few cases when ATP earned by respiration is utilized directly as the energy source of the streaming. The prerequisite for this is that energy-rich phosphate, which is generated in the mitochondria, takes the form of ATP in most cases and diffuses out from them into the ground plasm. It follows, then, that the extent to which the respiration energy can be utilized for the movement depends upon the diffusibility of ATP from mitochondria to the seat of motive force generation. Needless to say that this also depends upon the supply of inorganic phosphorus from the ground plasm to the mitochondria. Even though the level of ATP in the cell as a whole is raised through respiration, its utility for the protoplasmic movement is small if ATP cannot diffuse easily in the hyaloplasm.

Lettré (1951, 1954) and his collaborators (Lettré, Albrecht and Lettré 1951 a, b) have found the very important and interesting facts that, by bringing tissue-cultured fibroblasts under anaerobic condition or treating them with respiratory poisons, a remarkable protoplasmic movement can be induced in the resting cell such as is only observable at cell division. They interpreted this fact by saying that the undivided cells take the spindle form and are almost quiescent under aerobic conditions because they are in a state of continuous contraction owing to the high ATP level, whereas, if respiration is hindered, this high level of ATP gets lower so that the state of contraction is partly loosened and movement is produced. It is a significant fact in the physiology of protoplasmic motion including streaming that the motion is induced not by the rise in ATP level but by its local lowering. Lettré *et al.* came to this conclusion without taking into consideration the problem of the availability of ATP mentioned above. It may be that ATP synthesized in mitochondria readily diffuses in the case of the fibroblast. But what is left to be explained is why glycolysis poisons, such as iodoacetic acid, sodium fluoride, or an uncoupler of oxidative phosphorylation, such as 2, 4-dinitrophenol, do not induce the movement.

A possible alternative explanation may be that the movement gets more active under anaerobic conditions or through respiration poisons, not because the $\sim$ P level gets lower so as to release the cell from continuous contraction, but because the level of $\sim$ P that can be directly utilized for the movement gets higher probably on account of the Pasteur effect. Such a view may not only seem to be inconvenient for the experimental results obtained by Lettré *et al.* themselves, but also stands in conformity with the results we obtained with the slime mould, and moreover agrees well with the experiments of Goldacre and Lorch (1950) who microinjected ATP in the amoeba. Furthermore it agrees with the results obtained in a glycerinated 'cell model' studied by HOFFMANN-BERLING (1954 a, b, 1955).

As to the mechanism of the protoplasmic streaming being induced in the plant cell, there are still no detailed data, but it may also be reduced to the problem of the shift of the metabolic passway and that of the availability of ATP for the streaming.

All in all, the information gathered from the slime mould shows that what represents the direct energy source of the protoplasmic movement is not the mean value of the ~P level in the cell as a whole, but is the level of ~ P at the place where the mechanochemical system exists, or at the seat of the generation of the motive force. If the ATP concentration there gets higher, the movement gets generally more active up to the optimum level. In any event, as the glycolytic process and the generation of the motive force of the protoplasmic movement occur in the ground milieu (hyaloplasm) of the cytoplasm, the energy obtained by the glycolytic process can be readily utilized for the movement; the energy obtained by respiration acquires its significance as an energy source for the movement only when ATP is readily diffusible from mitochondria.

VII. Influences of External Factors

1. Temperature

a) Relation between temperature and rate of streaming

The effect of temperature on protoplasmic streaming has been studied for many years, going back as early as 1774 when the streaming was first discovered by CORTI. It was, however, NÄGELI (1860) who first studied in detail the influence of temperature on the rate of protoplasmic flow on a quantitative basis. His data show that the rate of flow in the *Nitella* cell increases according to an exponential function of temperature.

On the other hand, VELTEN (1876 a), who studied the temperature effect on the speed of streaming in *Chara, Vallisneria,* and *Elodea,* found that the streaming increases linearly with temperature. The results obtained by LAMBERS (1925, 1926) in *Characeae* (*Chara foetida, Nitella mucronata* and *N. translucens*) show also the linearity of the temperature-velocity relation. LAMBERS' experiment further makes it clear that the rate in *Nitella* always has a constant value with constant temperature independently of whether the cell has been heated to that point from a lower temperature, or whether it has been cooled down to that point from a higher temperature. This means that there is no indication of "hysteresis."

ZURZYCKI (1951) studied anew the temperature effect on the streaming in the *Elodea* leaf. Q_5 calculated from the data of his experiment are rather irregular, but the data fit satisfactorily into BĚLEHRÁDEK's formulation, $V = \frac{t^b}{a}$[4]. The results obtained by ZURZYCKI are also in accordance with the

[4] V: rate, t: temperature, a und b: constants (cf. BĚLEHRÁDEK 1929, 1935); in ZURZYCKI's data, $a = 1.95$ and $b \doteq 0.81$.

those of BEIKIRCH (1925), who studied the rate of streaming in *Elodea* and *Vallisneria* in relation to temperature.

In order to survey the results obtained by different authors in different materials and to facilitate mutual comparison, some of them are redrawn and brought together in Fig. 49.

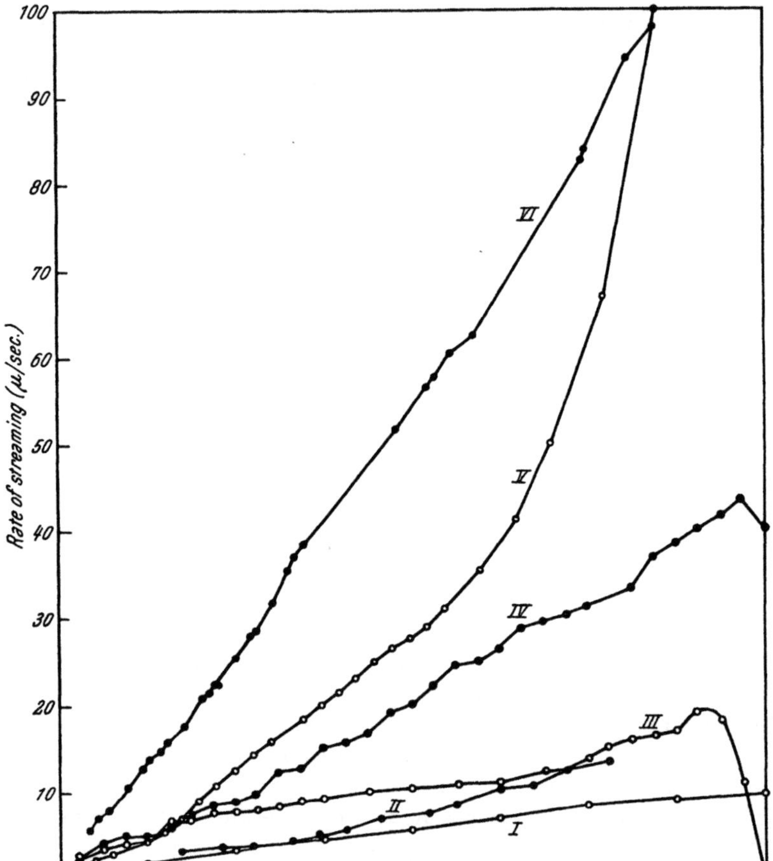

Fig. 49. Comparison of the results obtained by previous authors on the relations between temperature and the rate of protoplasmic streaming in different materials. Curve I: *Elodea densa* (ZURZYCKI 1951), curve II: *Oryza sativa* (root hair) (DOI 1950), curve III: *Elodea canadensis* (VELTEN 1876a), curve IV: *Vallisneria spiralis* (VELTEN 1876a), curve V: *Nitella syncarpa* (NÄGELI 1860), curve VI: *Nitella mucronata* (LAMBERS 1925).

The results of experiments thus far published may be classified in the following three major groups according to the mode of dependence of the rate of flow on temperature. 1. Linear relation [VELTEN 1876 a (*Chara, Elodea, Vallisneria*), LAMBERS 1925, 1926 (*Nitella, Chara, Tolypella*), YAMAHA and IDA 1933 b (*Nitella*), UMRATH 1934 (*Nitella*), etc.]. 2. Relation represented by exponential functions of temperature, including VAN 'T HOFF's formula and

ARRHENIUS formula[5] [NÄGELI 1860 (*Nitella*), BOTTELIER 1934, 1935 (*Avena*, 90 hours old coleoptile: $Q_{10} = 1.8$ within 7°–17° C., 140–200 hours old coleoptile: $Q_{10} = 1.8$ within 7°–25° C.), EYMERS and BOTTELIER 1937 (*Avena* cf. Fig. 50), DOI 1950 (root hairs of *Oryza, Triticum, Zea, Lycopersicum, Cucumis, Raphanus*), etc.]. 3. Relation represented in terms of BĚLEHRADEK's empirical formula, $V = \frac{t^b}{a}$ [BEIKIRCH 1925 (*Elodea, Vallisneria*), ZURZYCKI 1951 (*Elodea*)]. The linear relation is a special case of this formula when $b = 1$.

Whether the relation is linear or exponential or expressed by BĚLEHRADEK's formulation, depends upon the temperature range. Even when the relation is exponential, it can be regarded as linear in many cases, if the temperature in question is restricted to a narrow range.

ZURZYCKI (1951) distinguished two groups of plants which may be different in the temperature-velocity relation. Namely, he thinks that in green cells having chloroplasts, the relation obeys BĚLEHRADEK's formulation (BEIKIRCH 1925, ZURZYCKI 1951) or is linear (LAMBERS 1925, UMRATH 1934), while in cells having no chloroplasts, or in green plant cells in darkness where assimilation is interrupted, this relation is more likely to be represented by VAN 'T HOFF's law.

b) Factors participating in the effect of temperature on the streaming rate

BOTTELIER (1934, 1935) made a very thorough study of the influence of temperature upon the protoplasmic streaming in the cell of the coleoptile of *Avena*. He showed that the Q_5 of protoplasmic streaming in *Avena* is greater at lower temperature than at higher temperature, and that a younger coleoptile has a smaller temperature coefficient than an older one; At 3°–33° C., Q_5 is 1.05 in a cell of young coleoptile while it is 1.33 in a cell of an old coleoptile (Fig. 50). BOTTELIER (1935) and EYMERS and BOTTELIER (1937) interpreted this in terms of two factors: first, decrease in solubility of oxygen in the surrounding medium with the rise in temperature, and second, difference in oxygen-demand between younger and older coleoptiles for maintaining the streaming. When air is replaced by nitrogen, the temperature range where the temperature coefficient is small ($Q_5 = 1.05$) gets wider. When it is replaced with oxygen, the Q_5 becomes greater. These facts clearly show that the supply of oxygen is the limiting factor in the range where the temperature coefficient is small. In other words, the rate of protoplasmic streaming in the coleoptile of *Avena* is determined by age and the amount of oxygen. When the cell grows, the quantity of protoplasm becomes constant while the surface of the cell increases so that the amount of oxygen absorption is enhanced and the Q_5 gets greater. When the temperature is raised, solubility of oxygen in the surrounding medium diminishes with the result that the concentration of oxygen to be absorbed by the cell becomes less. Thus in a coleoptile of

[5] YAMAHA and IDA (1933 b) give the temperature characteristic (μ) of the streaming rate as follows: $\mu = 8450$ (*Chara*), 10,300 (*Nitella*), 8450 (*Vallisneria*), 4780 (*Elodea*).

medium age, Q_5 is large ($Q_5 = 1.33$) at a low temperature and small at a high temperature ($Q_5 = 1.05$).

DOI's experiments performed on various crop plants also show that the measured value is slightly smaller around 30° C. than the value expected on the basis of VAN 'T HOFF's law. This implies that Q_5 becomes somewhat smaller as the temperature rises. This is probably due to the limitation of O_2 supply, which is caused on one hand by decrease in O_2 solubility as pointed out by BOTTELIER and on the other by increase in respiration rate at higher temperature.

This is further supported by the fact that the maximum rate of flow in a root hair of *Oryza* is attained at 33° C. under ordinary conditions, whereas this optimum temperature is raised to 34° C. if the material is in a diluted H_2O_2 solution and is lowered to 31° C. in a diluted KCN solution. In case the material is submerged in well water for 16 hours the optimum temperature of streaming becomes 30°–31° C. These results show that under favourable conditions for respiration the optimum temperature for streaming is higher than under conditions inhibiting respiration.

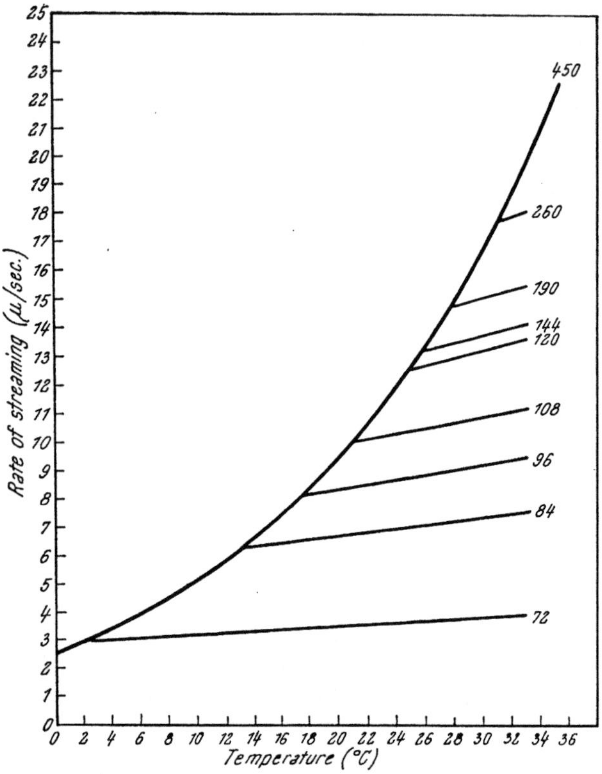

Fig. 50. Influence of temperature on the velocity of protoplasmic streaming. The figures at each curve indicate the age of the coleoptiles in hours. Van 't Hoff curves with $Q_{\frac{18}{23}} = 1.33$ and 1.05 respectively.
(EYMERS and BOTTELIER 1937.)

ZURZYCKI's work gives further information about the influence of temperature on the streaming when the oxygen comes only from one source—only from assimilation or only from diffusion from the environement. When oxygen is supplied by way of assimilation, the velocity-temperature relation follows the BĚLEHRÁDEK's formula, but when oxygen is supplied by way of diffusion, the cell being placed in darkness, the Q_5 coefficients are nearly constant (between 5° C.–30° C.) and the relation fits nicely into VAN 'T HOFF's formula.

According to ZURZYCKI (1951), the factor limiting the streaming rate when BĚLEHRÁDEK's law holds is probably the diffusion of oxygen liberated in the protoplasm in the course of photosynthesis. Another factor which might play a role is the possible diffusion from chloroplasts of ATP produced as a result of photosynthetic phosphorylation (ARNON 1955). In any event, when the *Elodea* cell is in darkness, the diffusion from without would be a limiting factor only in higher temperature just as is the case in the *Avena* coleoptile (BOTTELIER 1935). On the other hand, in lower temperatures, where VAN 'T HOFF's law holds, some other process—possibly the respiration of the plant—would be the limiting factor.

There are many experimental data concerning the temperature-velocity relation, but its analysis is always difficult and equivocal. LAMBERS (1925, 1926), on the ground of his data showing linear temperature-velocity relation, which reminded him of fluidity graphs, thought that the rate-determining process is the change in viscosity of the protoplasm. According to him, the increased rate of streaming at higher temperatures is due to a progressive increase in fluidity, *i. e.*, to a progressive decrease in viscosity. On the other hand, ROMIJN (1931) maintained that the viscosity of *Nitella* plasm finds its maximum around 20°–22° C. He drew this conclusion by means of the plasmolysis method (cf. WEBER 1924); whether temperature is heightened or lowered from 20° C., the plasmolysed protoplast tends to assume a more concave form. Obviously this conclusion does not agree with LAMBERS' interpretation. ROMIJN's conclusion, however, may also be not decisive, since, what actually takes part in the streaming is the endoplasm, whereas the viscosity estimated from the plasmolysis form is in the main dependent on the viscosity of the ectoplasm rather than on that of the endoplasm, or else on the adhesion of the protoplast to the cell wall.

The attempt to interpret the temperature effect on the speed of protoplasmic streaming has been confronted time and again with the problem as to how temperature affects the motive force apart from its effect on viscosity, a crucial problem for the causal analysis of the change in the rate of protoplasmic streaming. Experiments conducted on the myxomycete plasmodium throw some light on this problem.

c) The effect of temperature on the motive force of the protoplasmic streaming

As the double-chamber method (p. 40) enables us to measure the motive force responsible for the protoplasmic streaming in a slime mould, KAMIYA (1953) investigated how the force is affected by changes in temperature. For this purpose, the double-chamber was submerged in a small thermostat bath under the microscope. Measurement of the motive force was conducted at different temperatures with the interval of 5° C., *i. e.*, at 10° C., 15° C., 20° C., 25° C. and in some cases also at 30° C. after the plasmodium has been left at a new temperature for 30–60 minutes.

Fig. 51 shows the motive force curves (dynamoplasmograms, DPG) of one and the same plasmodium at 10° C., 20° C. and 25° C. From these curves it is clearly seen that the frequency of the motive force rhythm increases (the period decreases) as the temperature is heightened, but the amplitude of the DPG does not increase; on the contrary, it has a tendency rather to decrease as the temperature rises. From a series of experiments of this kind, values of the temperature coefficient were obtained in terms of Q_{10} both in respect to frequency and strength [6] of the motive force. They are summarized in Table 3.

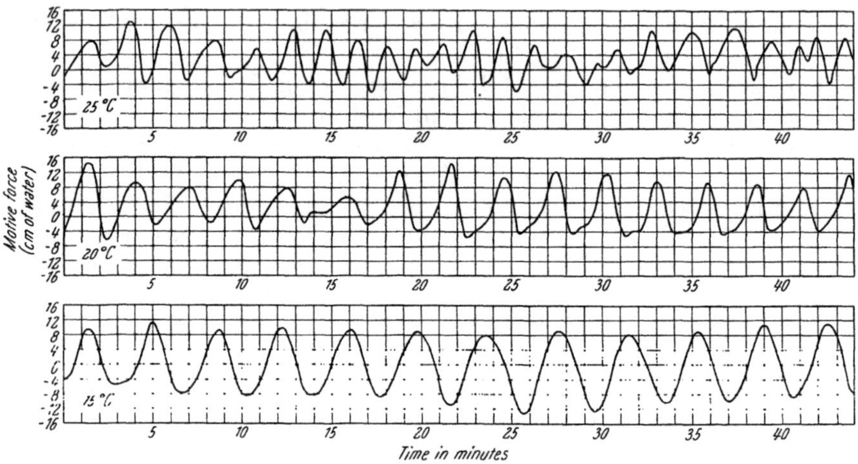

Fig. 51. DPG of one and the same plasmodium at different temperatures. (KAMIYA 1953.)

Table 3. *Q_{10} of the motive force of the protoplasmic streaming in plasmodia of Physarum polycephalum.*

Temperature range	10⁰–15⁰ C	15⁰–20⁰ C	20⁰–25⁰ C
Frequency	2.6	1.8	1.6
Strength	1.0	0.8	0.8

The very low Q_{10} values for the motive force is worthy of note. The temperature coefficients (Q_{10}) of the frequency (reciprocal of the period) coincide with those of the rate of protoplasmic streaming obtained in *Nitella* by LAMBERS (1925) (cf. KAMIYA 1953). In other words, the relation between the pulsation and temperature is approximately linear in the temperature range of 10⁰–25⁰ C. That temperature has little effect on the magnitude of the motive force responsible for the streaming is further supported by the fact that even when the two compartments of the double-chamber are brought to different temperatures, say one at 15⁰ C. and the other at 25⁰ C., the polarity line of the dynamoplasmogram shows no appreciable change (HATANO unpublished). This means that the absolute inner pressure of the plasmodium is not much affected by temperature.

[6] The strength of the motive force was calculated statistically in the following way: The areas included by the curve and the polarity line (p. 46) are summed up arithmetically both on the upper and lower sides of the polarity line. This total sum divided by the corresponding duration of time was taken as the measure of the mean strength of the motive force.

d) Supraoptimal temperature

It is obvious that there is generally a higher limit of temperature at which the streaming can maintain a constant speed. This is the temperature which we may call optimal. In many plant cells the optimal rate of the streaming is usually found in the temperature range between 27° C. and 37° C. If we raise the temperature still higher, the speed is indeed increased, but this is invariably followed by a spontaneous decrease eventually leading to a complete stop. LAMBERS gave a curve (Fig. 49, VI) showing that streaming rate increases linearly with temperature in *Nitella mucronata* between 2° C. (8μ/sec.) and 34° C. (100 μ/sec.). Above 34° C. the rate of flow still increases, but the rate depends on the duration of heating.

Using internodal cells of *Chara*, YAMAHA and IDA (1933 b) studied with special precaution how the speed of flow changes with time at supraoptimal temperature range. Above 40° C. the flow is increasingly retarded at a constant temperature eventually coming to a standstill; the durations before the flow stops are 120, 45, 30, 20, and 10 minutes at 40°, 42°, 44°, 45°, and 46° C., respectively. At 40° C., the protoplasm undergoes reversible thermal gelation periodically, accompanied by temporary and uneven decrease in the streaming rate.

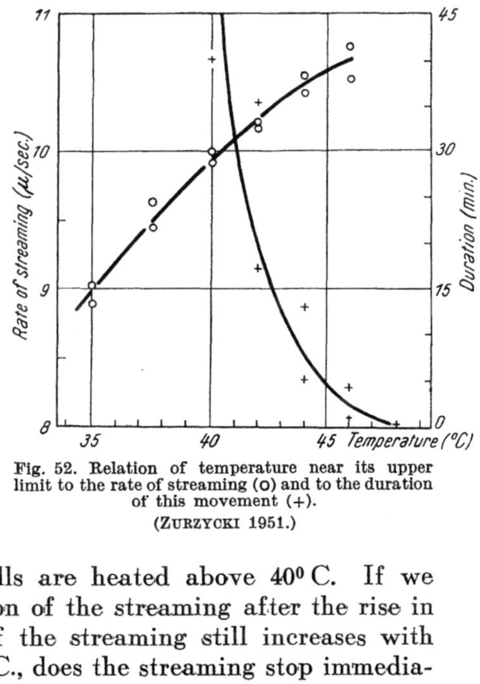

Fig. 52. Relation of temperature near its upper limit to the rate of streaming (o) and to the duration of this movement (+). (ZURZYCKI 1951.)

In *Elodea* cells, too, according to ZURZYCKI (1951), the protoplasmic streaming stops sooner or later if cells are heated above 40° C. If we confine our attention to the acceleration of the streaming after the rise in temperature, we find that the rate of the streaming still increases with temperatures up to 46° C.; only at 48° C., does the streaming stop immediately. In the supraoptimal range, therefore, the temperature effect should be considered in terms not only of an instantaneous speed of flow, but of how long the streaming is maintained at that temperature. Fig. 52 shows the temperature-velocity relation at the proximity of the upper limit of the physiological range and the duration in which flow is maintained. From this figure we understand that the optimal temperature in *Elodea* is less than 40° C.

e) Sudden change in temperature

It has been repeatedly reported since the last century that the rate of protoplasmic rotation shows a peculiar transient change, when the temperature is raised or lowered suddenly. In case of a sudden lowering of temperature, the flow often stops instantly as a result of shock. It is probably DUTROCHET (1837) who first studied the response of protoplasmic streaming to a sudden change in temperature. He found in *Chara flexilis* that, both when raising and lowering the temperature, the streaming does not attain immediately the rate characteristic of the new temperature, but undergoes a transitory retardation. HOFMEISTER (1867) also obtained similar results. With sudden cooling from 18.5° C. down to 5° C. he observed a temporary stoppage of the streaming. He also recognized the same phenomenon when the temperature is raised suddenly, *e. g.*, from 18° C. to 27° C., from 27° C. to 34° C. or from 34° C. to 40° C. These observations were confirmed by HARVEY (1942) in *Nitella flexilis.* There must be always a certain lapse of time before the cell recovers from this shock effect.

The results of the experiments were, however, not always concordant. VELTEN (1876 a) found no transient retardation in *Chara, Nitella, Elodea* and *Vallisneria,* no matter whether the temperature was raised or lowered. HÖRMANN (1898), on the other hand, found a temporary retardation only when the temperature was lowered, but not when it was raised. Many other subsequent investigators obtained a similar result (LAMBERS 1925, 1926; COOK 1929; ROMIJN 1931; etc.).

The effect of a sudden change in temperature has been studied most thoroughly by COOK (1929). COOK (1929) transferred cells of *Nitella clavata* from 20° C. to 35° C., but there was no cessation of streaming. There was of course a change of rate from that characteristic of 20° C. to that of 35° C. If the cell is transferred from 35° C. to 20° C., the flow continues without interruption but at a lower rate, *i. e.,* the rate characteristic of 20° C. Once, however, the cell is brought from 35° C. to water at 10° C. instead of 20° C., the flow ceases suddenly and does not recommence for at least 2 minutes. This abrupt stop is a phenomenon distinguished from the mere change of speed observed between 35° C. and 20° C. In order to bring about complete stoppage the exposure to the low temperature (10° C.) need not be longer than half a second. Obviously, the sudden transition from the high to the low temperature is essential as a stimulus, and a prolonged exposure at the low temperature is not needed to bring protoplasm to a temporary standstill. The stoppage of streaming does not occur at a definite point on the temperature scale but is rather a function of the speed of change in temperature. By transferring cells from 20° C. to 1° C. COOK (1929) found the interesting fact that "the longer the exposure to the low temperature the more rapid the recovery when the cell is replaced at the higher temperature." COOK assumed that recovery is proceeding at 1° C. even though there is no motion perceptible. Thus he says: "Recovery of the *power to stream* can take place while there is no *actual streaming* in progress," and comes to the conclusion that "the two factors, the actual streaming and the capacity for streaming, must be entirely separate and distinct."

ZURZYCKI (1951) found in *Elodea* cells that when the temperature was dropped suddenly, a temporary decrease of the rate or interruption of the streaming in *Elodea* cells occurs, which depends both on the temperature difference and on the new temperature. The change in rate is transient, and after 10–20 minutes the rate assumes its normal value characteristic of the new temperature. When the temperature is suddenly increased, the rapidity usually increases immediatly to the value characteristic of the new temperature. In some cases a small and very temporary increase above this normal value is noted (Fig. 53). Such an "overshoot" phenomenon caused by a sudden rise in temperature has not been reported before. In the case of a sudden cooling, ZURZYCKI's observation is in accordance with the results obtained by previous investigators.

We are not yet quite sure if such a transient change in the rate of streaming caused by a rapid change in temperature can occur independently of electrical response of the cell. Therefore one of the important points to be dealt with in respect to the temperature shock is to determine whether

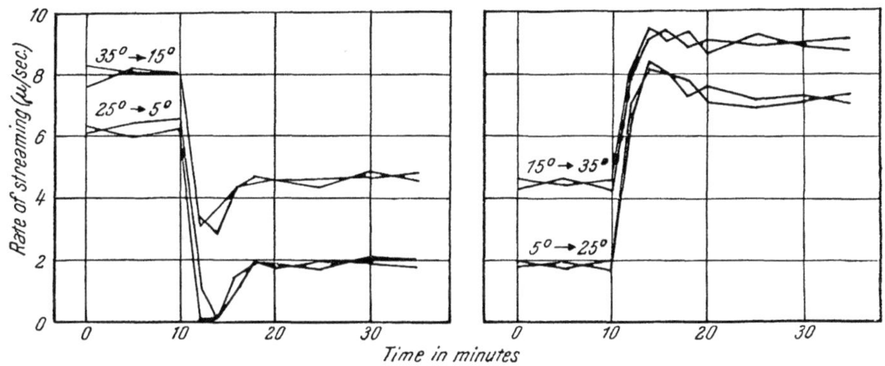

Fig. 53. Changes in rapidity of protoplasmic streaming caused by sudden cooling (left) and sudden heating (right).
(ZURZYCKI 1951.)

or not the sudden retardation or standstill of the streaming is accompanied by the generation of an action current.

2. Light

It has been known that visible light can retard, stop or accelerate protoplasmic streaming. Sometimes it can also induce streaming in quiet cells. Arbitrarily, we shall consider the effect of light from two different aspects; first, the effect on the streaming protoplasm, and second, the effect initiating the streaming in previously quiescent cells, *i. e.*, photodinetic effect. For the time being we shall deal with the former. To the latter we shall come in a later chapter (p. 133).

The action of light on protoplasmic streaming has been the subject of study of many workers for a long time. HOFMEISTER (1867), REINKE (1871), PRINGSHEIM (1881), JOSING (1901), EWART (1903), etc. are some of the authors in classic literature. The old work was well surveyed by BOTTELIER (1934). Obviously, the light effect must be elucidated in terms of wave length, intensity, and duration of exposure. The necessity for eliminating the heat effect is self-evident. Much of the old literature is not satisfactory in these respects. In view of the fact that the rate of flow is under the direct influence of protoplasmic viscosity, and that the viscosity is very susceptible to light, it is well to be expected, if only from these facts, that the streaming rate must necessarily be modified through light.

Bottelier (1933, 1934) made an elaborate series of experiments on the effect of light on the protoplasmic streaming in coleoptile cells of *Avena*. He found that a brief illumination with blue light after a period of 3–4 minutes evokes a retardation of flow lasting about 4 minutes. The reaction depends upon the amount of light energy admitted. When the light energy

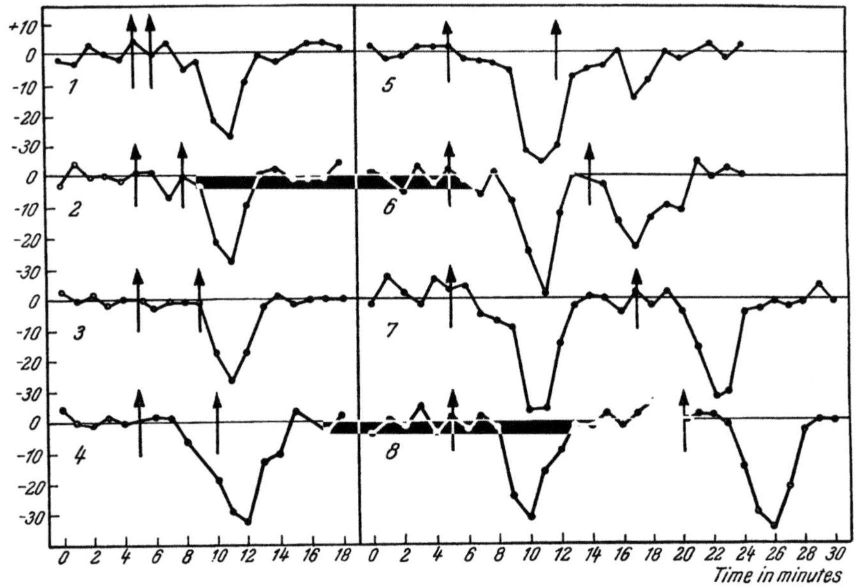

Fig. 54. Retarding effect on the protoplasmic streaming in *Avena* coleoptile caused by two successive exposures with weak light (2×190 erg/cm²) in different intervals. Arrow = Beginning of illumination. Abscissas: time in minutes. Ordinates: deviation in per cent from the mean velocity. Duration between two illuminations. — No. *1*: 1 min., No. *2*: 3 min., No. *3*: 4 min., No. *4*: 5 min., No. *5*: 7 min., No. *6*: 9 min., No. *7*: 12 min., No. *8*: 15 min.

(Bottelier 1934.)

is small, retardation follows. If the cell is exposed to a great light quantity (800 ergs/cm²), acceleration of flow ensues. The degree of reaction is solely dependent upon the amount of supplied energy and obeys the product law.

When the cell is illuminated twice repetitively, with an interval of less than 3–4 minutes, the reaction appears to correspond to the total light quantity, provided that the second illumination is admitted before the retardation appears that is caused by the first illumination (Fig. 54, Nos. 1–3). If the interval amounts to 10 minutes or more, then two equal reactions are brought about (Fig. 54, Nos. 7 and 8). If the cell is secondarily illuminated while the streaming is retarded under the effect of the first illumination, that is, during the time interval of 4–8 minutes after the first illumination, the second reaction appears smaller than the first (Fig. 54, Nos. 4–6). The continuous illumination induces a strong reaction 4 minutes after the beginning of illumination, which however usually disappears again.

As for the spectral sensitivity of the streaming in *Avena* coleoptile, blue light is most effective, then violet and ultraviolet; the effect of green light and light of longer wave lengths is negligible (Table 4). This type of spectral sensitivity indicates that energy absorbed by yellow pigments is decisive, and coincides well with that of the photogrowth reaction of the coleoptile of *Avena*. It is possible, then, that the retardation of the streaming is brought about by way of a photoreceptor which is the same as that for the photogrowth reaction. In both cases there are similar relations of the intensity and the duration of stimulus to the two reactions. For the inducement of the protoplasmic streaming through light (photodinesis), however, red light is most effective, then blue and green (p. 133), which shows that the photodinetic effect is brought about through the mediation of chlorophyll acting as sensitizer.

Table 4. *Sensitivity of the protoplasmic streaming in Avena coleoptile for light of different wave lengths (after* Bottelier *1934).*

Wave length (Å)	Sensitivity relative to blue light (4360 Å)
3660 (ultraviolet)	12
4050 (violet)	50
4360 (blue)	100
5460 (green)	1.7
5780 (yellow)	<1
6200 (red)	<1

The effect of light on protoplasmic streaming is noticeable only in the cells which are illuminated directly. There is no sign of propagation of the stimulus evoked through visible light. Virgin's observation (1950, 1951) on viscosity shows that its change is restricted to the very locus which is exposed to light, the nearby region kept in darkness being unaffected. In the case of ultraviolet radiation, as will be stated later (p. 134), the situation is different.

As the effect of light on protoplasmic streaming cannot be considered independently of its effect on viscosity, it will be pertinent at this point to touch upon the light effect on protoplasmic viscosity. It has been observed repeatedly that the viscosity of the protoplasm is changed upon illumination. Weber (1929) found that the plasmolysis time for the illuminated leaves of *Ranunculus ficaria* is longer than that for the control in darkness. Later the effect was studied in *Elodea* cells by Stålfelt and Virgin. The viscosity of protoplasm in *Elodea* cells kept in darkness shows a great sensitivity to light. Stålfelt (1945, 1946) observed a marked fluctuation in the viscosity of protoplasm in the leaf of *Elodea densa* upon illumination. Only under strong light (32,000—48,000 lux) the alteration in viscosity ceases, but the viscosity starts fluctuating again when light intensity is decreased. Photosynthesis may not be the cause, for illumination of 100 lux lasting only

15 seconds is sufficient to induce the change. VIRGIN (1947, 1950) found the threshold value to be as low as 1 lux and, in autumn even 0.001 lux. The effect on viscosity in the *Elodea* cell finds its maximum in the range of 4600–4800 Å, the effect rapidly decreasing toward longer wave lengths. It is interesting that this nearly coincides with the spectral sensitivity of the protoplasmic streaming in *Avena*.

Protoplasmic streaming in *Elodea* leaf is stopped when the supply of oxygen is cut off. As mentioned already (pp. 66 and 84), ZURZYCKI (1951) observed that the streaming starts again and is kept at a steady rate upon illumination, even when no oxygen is available from the outside. Obviously, an effect of this type is mediated by photosynthesis. A point which is worth noticing in ZURZYCKI's result is that the rate of flow is greater in the presence of light without outside oxygen than in the presence of extracellular oxygen without photosynthesis. Whether this difference is due to the viscosity change caused by light or to the difference in the mode of oxygen supply is unknown. In this connection the possibility should not be overlooked that ATP synthesized in chloroplasts by way of photosynthetic phosphorylation (ARNON 1955) is utilized directly as the energy source for the protoplasmic streaming.

An interpretation of the light effect and its causal connection with various theories on the mechanism of protoplasmic streaming is so far impossible.

3. Ultraviolet rays and other radiations

The effect of ultraviolet rays may be considered in terms of two different aspects just as in the case of visible light, *i. e.*, alteration of the behaviour of flowing protoplasm and initiation of the flow in quiescent cells. We shall deal with the former here.

Among early workers in this field were HERTEL (1904), SCHULZE (1910), VOUK (1913), etc. Using ultraviolet rays with a wave length of 2800 Å, which coincides with an absorption maximum of proteins, HERTEL observed the effect on the *Elodea* cell, SCHULZE on staminal hair cells of *Tradescantia*, root hairs of *Hydrocharis* and leaf cells of *Vallisneria*. They observed that the streaming is gradually retarded and is eventually led to a stop. VOUK's observation on myxomycete plasmodia shows also that the protoplasm ceases to flow after a few minutes' exposure to the mercury lamp, the cessation being mostly reversible.

NOETHLING and ROCHLIN (1931) exposed an *Elodea* leaf to ultraviolet rays (with wave lengths longer than 2800 Å) of a constant energy for unequal durations of time from 2 up to 60 minutes and measured the rate of movement of chloroplasts 5 hours after the beginning of the exposure. It was shown in Fig. 55 that the rate increases with exposure time at first, but it begins to decrease if the dose of ultraviolet ray is further increased. With 60 minutes' exposure, no movement is observable and there is a beginning of necrosis. GIBBS (1926) reported that radiation of 2378–3126 Å causes first liquefaction and then, on longer exposure, coagulation of the protoplasm of *Spirogyra*. HEILBRUNN (1928, 1940, 1952) stressed that calcium plays a significant part in the changes in protoplasmic viscosity, either in its increase or decrease, under the influence of ultraviolet rays.

Another interpretation accounting for the initial increase in the rate brought about by appropriate doses is that substances effective in stimulating the streaming have been produced as a result of the decomposition of cell constituents. These substances may possibly be amino acids, which are potent inducers of the protoplasmic streaming (p. 130). The effect of ultraviolet rays in slowing down the protoplasmic streaming and leading to eventual stoppage is most probably brought about through their action on the viscosity of protoplasm.

HARVEY (1942 a) observed that a high intensity flash of ultraviolet light frequently stops the protoplasmic streaming in *Nitella flexilis* at the upper surface of the cell where the ultraviolet light enters, but not below. No effect is observable if the cell is covered with 1.1 mm. glass, even under an intense flash (22,000 volts), but rotation stops immediately if 0.5 mm. of quartz is substituted for the glass. Sometimes the stoppage is permanent, which means necrosis, but frequently the effect is reversible and the cell eventually resumes its original rate of streaming.

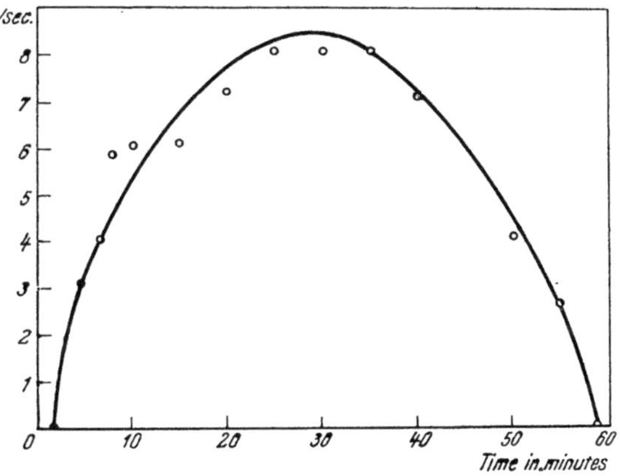

Fig. 55. Dependence of the rate of protoplasmic flow (*Elodea*) upon the time of exposure to ultraviolet rays. (NOETHLING and ROCHLIN 1931.)

A flash of ultraviolet light behaves, depending on its intensity, just as an electrical stimulus which may abruptly stop or slow down the streaming in *Nitella*. As a matter of fact, HARVEY found that ultraviolet stimulation is often accompanied by an action potential, which may or may not be propagated as an excitation wave. If the stimulus is weak, no action potential is evoked. In such a case, the streaming slows down gradually until it stops. The gradual cessation is distinct in its cause from the shock stoppage accompanying the action potential.

Pathological changes brought forth in myxomycete plasmodia through exposure to ultraviolet rays were studied by VOUK (1913) in *Didymium* and *Chondrioderma* and by SEIFRIZ (1939 b) in *Physarum*. When the plasmodium of *Fuligo septica* is subjected to ultraviolet rays, local small injuries are produced, which are distinguished from uninjured regions by a red colour (LOCQUIN 1949). It is a characteristic of the *Fuligo* plasmodium that it turns red when it is injured by mechanical means. This is presumably due to the activity of some oxidase in the plasmodium.

The effect of X-ray and radium was also a subject of study of some investigators in connection with the protoplasmic streaming. Already

60 years ago, Lopriore (1897) observed that small doses of X-ray accelerate the protoplasmic streaming in the *Vallisneria* cell. Williams (1923), using the stalk cells of *Saxifraga umbrosa*, showed that short exposures increase the rate of protoplasmic streaming as well as the Brownian movement in the protoplasm. Longer exposures depress the rate, and eventually stop the flow completely. Nadson and Rochlin (1934) obtained similar results in *Allium* epidermis and found that the period of acceleration in circulation is followed by retardation.

Wanner (1945) found in root hairs of *Hydrocharis morsus ranae* and epidermal cells of *Allium cepa* that there is an immediate decrease in the viscosity of the protoplasm in epidermal cells of *Allium* and an increase in the rate of streaming in the root hairs of *Hydrocharis* when these cells are subjected to 400–10,000 r. The increased rate is not closely correlated with the radiation dose. Up to 6,000 r, the original rate is resumed after a 10 hours' period of acceleration; with higher dosage, namely from 6,000 to 12,000 r, the eventual death of the cell ensues after the period of acceleration. If the dose exceeds 12,000 r, an immediate retardation of the streaming is brought about accompanied by deterioration of the protoplasm.

Using a much higher dosage, Bishop, McLaughlin and Tapley (1949) studied the effects of X radiation on the rate of cytoplasmic streaming in stamen hairs and pollen tubes of *Tradescantia paludosa* (diploid) and *Tradescantia virginiana* (tetraploid). No significant change in the rate of streaming of cytoplasm in either stamen hairs or pollen tubes was found with low doses of X rays. In both diploid and tetraploid stamen hairs, dosage as high as 700,000 r was required to bring all streaming to a standstill. However, since Brownian motion is evident up to 750,000 r, the cessation of flow is not accounted for as being due to viscosity increase. In the pollen tubes of diploid and tetraploid plants, which are more sensitive to X-rays, all streaming was stopped under 250,000 r and 200,000 r, respectively.

Williams (1925) studied the effect of radium on the protoplasm of *Saxifraga umbrosa*. In this case, too, a decrease in viscosity together with an increase in the rate of protoplasmic streaming is produced upon short exposure. Longer exposure to radium results in coagulation and vacuolization of the protoplasm. The effect of radium on the myxomycete plasmodium has been studied by Seifriz (1936) in *Physarum* and by Locquin (1949) in *Fuligo*. The observations of the two authors show that the tolerance of the plasmodia for radium radiations is exceedingly great. The latter author says that β-ray in proper dose accelerates the protoplasmic streaming. A transitory decrease of the protoplasmic viscosity (*Elodea*) under the effect of β- and γ-rays of radioactive isotopes was observed by Virgin and Ehrenberg (1953).

4. Osmotic pressure

Ewart (1903) found that no change in osmotic pressure is to be noticed when the streaming commences in a previously quiescent cell of *Elodea*,

Vallisneria, Trianea bogotensis, Aristolochia Sipho, Lepidium and *Tradescantia*. Thus EWART is of the opinion that the effect of osmotic pressure is insignificant if not negligible.

JURIŠIĆ (1922) confirmed EWART's statement in *Elodea* leaves by means of HÖFLER's plasmometric method (1918) and showed that cells have the same osmotic pressure no matter whether the protoplasm is quiescent or in active motion. Hence, it has been concluded that the so-called secondary streaming in plant cells bears no direct relationship to the osmotic pressure or to its changes. What these workers aimed at was to ascertain whether or not there may be a difference in osmotic pressure between the cells under quiescent and streaming state. Naturally such an experiment is possible only in a cell which exhibits a "secondary streaming" like *Elodea*.

Experiments attempting to elucidate the effect of osmotic pressure on protoplasmic streaming were not conducted for a long time until KURODA's recent work (1956). This is probably due to the fact that there has been no adequate method known to modify the osmotic pressure of a cell far above or below its normal level without bringing about alteration of the protoplasmic configuration, such as is the case when cells are thrown into plasmolysis. Using an internodal cell of *Nitella*, KAMIYA and KURODA (1956) developed a simple method to obtain cells having diverse osmotic pressures by tying off an internodal cell of *Nitella flexilis* with strips of silk thread after inducing transcellular osmosis (KAMIYA and TAZAWA 1956) (Fig. 56). By this means it is possible to produce cells, or exactly speaking, cell fragments, which have variable osmotic pressures within the range from as high as 3 times to as low as ¼ that of the normal cell. This increase or decrease in osmotic pressure is brought about either by condensation or dilution of the natural cell sap.

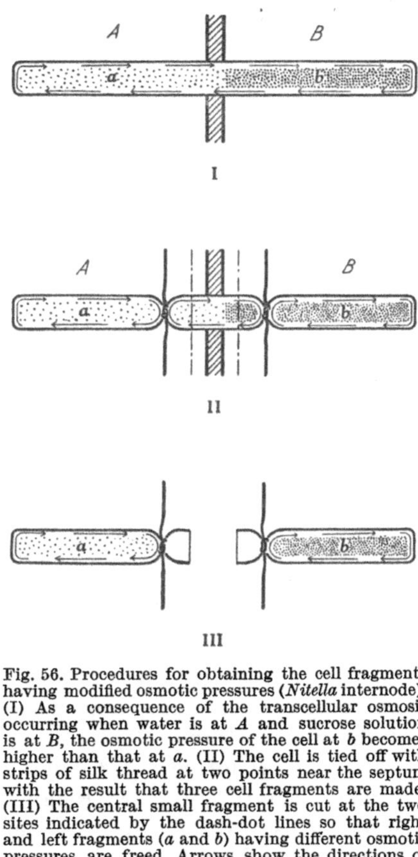

Fig. 56. Procedures for obtaining the cell fragments having modified osmotic pressures (*Nitella* internode). (I) As a consequence of the transcellular osmosis occurring when water is at *A* and sucrose solution is at *B*, the osmotic pressure of the cell at *b* becomes higher than that at *a*. (II) The cell is tied off with strips of silk thread at two points near the septum with the result that three cell fragments are made. (III) The central small fragment is cut at the two sites indicated by the dash-dot lines so that right and left fragments (*a* and *b*) having different osmotic pressures are freed. Arrows show the directions of the protoplasmic streaming. (KURODA 1956.)

The protoplasmic streaming in a cell fragment whose osmotic pressure has been made to deviate greatly from the normal level, which is equivalent to 0.26 M sucrose, settles at a new constant rate 30 minutes after the tying-off operation. But this settled rate is different from the rate in a cell whose osmotic pressure is normal. The rate of the streaming diminishes whether the osmotic pressure becomes higher or lower than the normal value (Fig. 57). At an osmotic pressure equivalent to 0.7 M sucrose or higher, the streaming stops completely. That this change in the rate is due solely to the abnormal concentration of the cell sap and not to the shock accompanying the operation is clear from the fact already referred to (p. 28) that ligation *per se* does not modify the rate of flow provided the fragment is longer than 1 mm. (HAYASHI 1952).

As for the viscosity of the protoplasm, which KURODA determined by means of a centrifugal method, it becomes lower than normal when the cell sap concentration is lower than normal, while it does not change much when the cell sap concentration gets higher than normal. Thus it follows from KURODA's experiments that the change in osmotic pressure of the cell causes diminution of the motive force responsible for the protoplasmic flow. As the alteration of the ionic concentration of the cell sap may in turn affect the ionic concentration of the protoplasm, we can assume from the above facts that ion concentration of the protoplasm under natural conditions is at its optimum for the generation of the motive force. It is likely that a proper concentration of certain ions is essential for the development of the motive force.

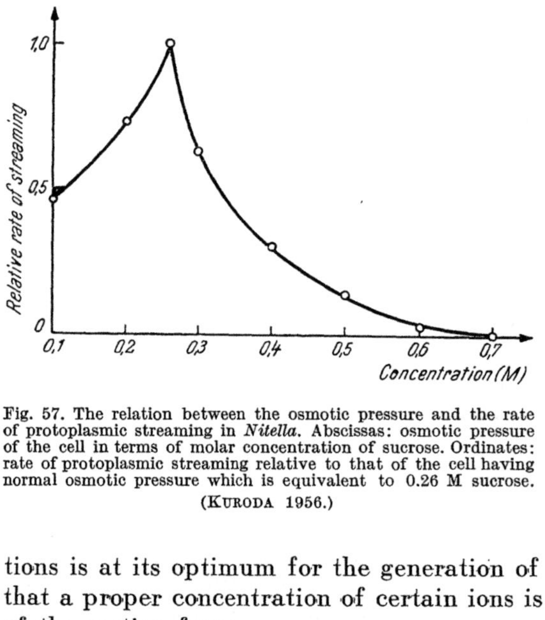

Fig. 57. The relation between the osmotic pressure and the rate of protoplasmic streaming in *Nitella*. Abscissas: osmotic pressure of the cell in terms of molar concentration of sucrose. Ordinates: rate of protoplasmic streaming relative to that of the cell having normal osmotic pressure which is equivalent to 0.26 M sucrose. (KURODA 1956.)

As was reported by KAMIYA and KURODA (1956), protoplasm with abnormal osmotic pressure regains its normal value spontaneously in the course of several days, thus showing marked ability of osmoregulation. Hand in hand with the restoration of the osmotic pressure to the normal level, the protoplasmic streaming also increases its rate, eventually reaching the normal value.

5. Plasmolysis

It has been known for nearly a century that protoplasmic streaming can continue in not a few cases (root hairs of *Hydrocharis*, stamen hairs of *Tradescantia*, leaf cells of *Elodea* or *Vallisneria*, etc.) even after the cell is subjected to plasmolysis (cf. HOFMEISTER 1867, VELTEN 1873, etc.). Admitting that there are also ample cases where streaming is abolished when the cell is thrown into plasmolysis, the capacity of protoplasm to flow in some plasmolysed cells proves that the contact between the protoplast and the cell wall is not an essential condition for the mechanism of the protoplasmic streaming. Protoplasmic streaming can take place even in a naked protoplast isolated from the cell after being plasmolysed. If a long cell, such as a long root hair of *Hydrocharis,* is plasmolysed, the cell contents are divided usually into several protoplasts in each of which rotation occurs in the same direction as before plasmolysis.

The plasmolysed cell is influenced by various factors: rise in the concentration of the cell sap, the chemical effect of a plasmolyticum (*e. g.*, dehydration or sometimes hydration of the protoplasm), mechanical sepa-

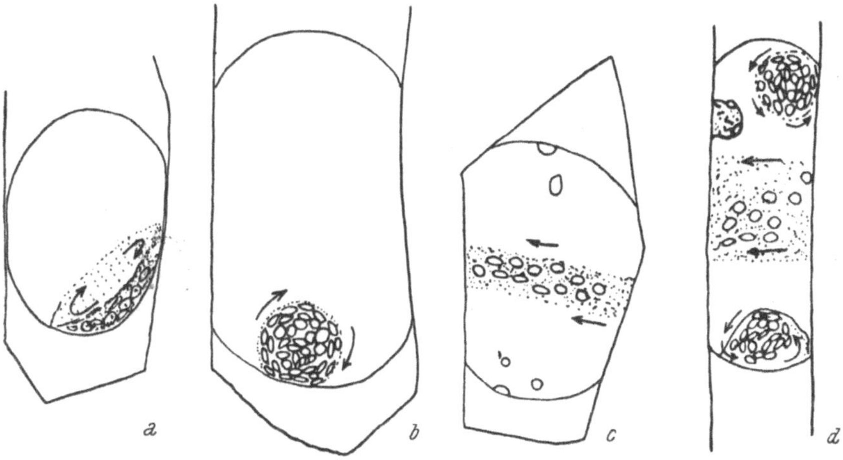

Fig. 58. Abnormal figures of protoplasmic streaming in plasmolysed cells of *Elodea densa*. *a*: The streaming protoplasm is gathered in the form of a disc which rotates as a mass around the central axis (profile), *b*: A similar condition seen from above, *c*: A band-like streaming zone, *d*: several lens-like protoplasmic masses and a band-like zone showing streaming.
(Küster 1910.)

ration of the protoplasts from the cell wall, changes in the form of the protoplast or in the distribution of the protoplasm, etc. Therefore the analysis of the effect is rather complicated. Generally, the rate of flow decreases to a considerable extent if the cell is plasmolysed. There are however exceptions. Sweeney (1944), for instance, found that the absolute rate of flow in plasmolysed root hairs of *Avena* remains unchanged for at least an hour after plasmolysis in 0.4 M fructose.

In a plasmolysed cell, protoplasm often suffers remarkable alteration in its distribution. Change in the distribution of protoplasm induced through plasmolysis leads to conspicuous abnormalities of the streaming figure as shown by Küster (1910). In Fig. 58 are shown several types of protoplasmic streaming in plasmolysed *Elodea* cells. Here we see protoplasm aggregating in the form of a lens, disc or belt which is in vigorous rotation. It is noticed that just a definite locus of protoplasm remains capable of rotating or frequently also many portions of protoplasm rotate independently of each other. The streaming continuing in the rotating disc shown in Fig. 58 *a* resembles what is called revolving motion by WIGAND (1885).

The internodal cell of *Nitella* or *Chara* is usually not readily plasmolysed, but still it is not impossible to keep the cell in a plasmolysed state in 0.24 M $Ca(NO_3)_2$ solution for more than 2 months. The streaming continues very actively in such a cell where we can observe many startling and instructive patterns of streaming (Fig. 59). Before plasmolysis the streaming course was nearly parallel to the longitudinal axis of the cell. When plasmolysis sets in, the structure of the cortical layer undergoes a pronounced change; chloroplasts agglomerate in the form of ring perpendicular to the cell axis or of lens, or in more irregular forms. HAYASHI and KAMITSUBO (1959) observed in detail this interesting phenomenon. According to KAMITSUBO (unpublished), the direction of streaming also changes as the

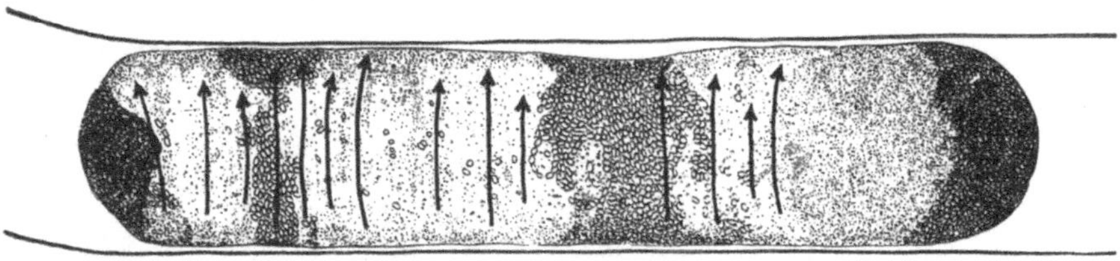

Fig. 59. A protoplast of *Chara Braunii* ("leaf" cell) plasmolysed with 0.24 M $Ca(NO_3)_2$ for the entire period of 70 days. The total length of the plasmolysed protoplast is about 600 μ; the width is about 130 μ. Streaming rate: 23 μ/sec.
(HAYASHI and KAMITSUBO, 1959.)

cortical gel is gradually directed at right angles to the axis. This phenomenon seems to correspond with the type of band-like streaming as observed by KÜSTER in *Elodea* (cf. Fig. 58 *c, d*). The reason why the direction of the streaming course is changed by plasmolysis has not been made clear as yet.

6. Hydrostatic pressure

Work conducted by BROWN (1934, etc.), MARSLAND (1936, 1938, etc.), EBBECKE (1936), HARVEY (1942 b), etc., has demonstrated that high hydrostatic pressure imposes solation upon the protoplasmic gels of animal and plant cells generally. The facts known up to 1940 are beautifully summarized by MARSLAND (1942).

It is particularly interesting and important in looking into the mechanism of protoplasmic flow to study the relationship between hydrostatic pressure, structural characteristics of protoplasmic gels and protoplasmic streaming. In the cells of *Elodea canadensis,* the rate of rotational streaming diminishes regularly with increasing hydrostatic pressure, complete cessation of flow occurring between 400 and 500 atmospheres. The effect is freely reversible if the material is not subjected to higher degrees of compression (above 400 atm.) for more than ½ hour. The slowing of flow occurs exactly in proportion to the degree of solation which increasing pressure induces in non-flowing parts of the protoplasm (Fig. 60).

PEASE (1940) studied the effect of high hydrostatic pressure upon the plasmodia of *Physarum polycephalum.* The result obtained is basically the same in that the plasmagel of the plasmodium tends to solate with increasing pressure. At a hydrostatic pressure above 270 atm. the flow is greatly retarded. The plasmagel forming channel walls is almost entirely solated so that the flow is primarily under the control of surface tension forces.

The retardation of the streaming by hydrostatic pressure in *Nitella* is not so regular as in *Elodea* (HARVEY 1942 b). Some cells show a progressive retardation of rotation as the pressure is increased in steps of 68 atm. while others manifest little change in rate until 270 atm. is reached after which progressive decrease occurs. At 60 atm. the rate is about ½ of that at ordinary atmospheric pressure. Protoplasmic streaming does not stop completely at 680 atm.

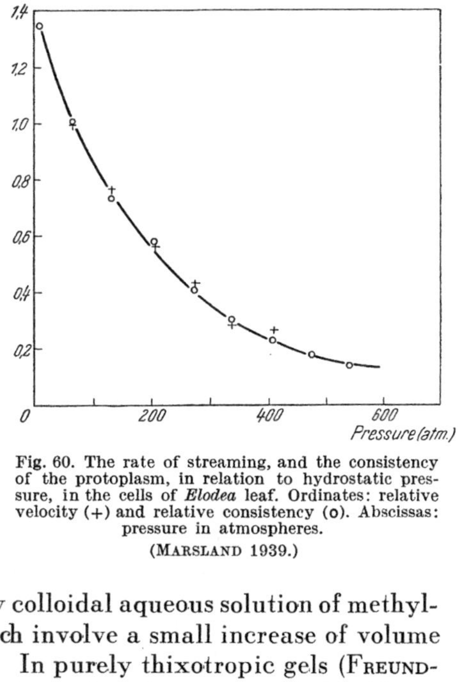

Fig. 60. The rate of streaming, and the consistency of the protoplasm, in relation to hydrostatic pressure, in the cells of *Elodea* leaf. Ordinates: relative velocity (+) and relative consistency (o). Abscissas: pressure in atmospheres. (MARSLAND 1939.)

It is known that high hydrostatic pressure markedly affects various metabolic reactions both as to rate and point of equilibrium. On the other hand, it is also a well-established fact that pressure can induce a pronounced change in sol-gel equilibria of certain colloids which is accompanied by a slight increase or decrease when gelation or solation occurs (HEYMANN 1935, 1936; FREUNDLICH 1937). In gels exemplified by gelatin or agar (FREUNDLICH's group II), which show a small decrease of volume with setting, pressure favours gelation. In gels represented by colloidal aqueous solution of methylcellulose (FREUNDLICH's group III), which involve a small increase of volume with setting, pressure favours solation. In purely thixotropic gels (FREUNDLICH's group I), in which sol-gel transformation is accompanied with no change in volume, pressure has little or no effect. Thus, if we regard the solation of protoplasm induced by pressure as being due to the direct action of pressure, protoplasm belongs to group III of FREUNDLICH's classification since all the protoplasmic gels so far studied by MARSLAND, BROWN, etc. proved to be highly susceptible to solation under pressure. MARSLAND and BROWN (1942) further showed that myosin (actomyosin) sol extracted from rabbit muscle behaves under pressure in the same manner as gels of "group III."

The question arises as to whether the pressure effect brought about on the living protoplasmic gel represents the result of a direct action upon its sol-gel equilibrium just like the pressure effect on the non-living gel belonging to FREUNDLICH's group III, or whether it is mediated by the shift of some metabolic equilibria in the cell, which in turn may tend to solate the protoplasm. MARSLAND (1942) says that the weight of evidence would seem to favour the direct action view.

Apart from the nature of the pressure effect, we are interested in the findings of MARSLAND and his associates since they contain important facts in considering the mechanism of protoplasmic streaming. A significant feature of the pressure effect is that the degree of retardation in the rate of protoplasmic flow and the degree of the solating action are numerically parallel. It would seem paradoxical that the streaming of protoplasm stops when it is solated. Here we realize through a concrete fact that a gel is essential for generating the motive force. And without the motive force protoplasmic streaming can never occur however low the viscosity may be.

7. Mechanical pressure

In experiments on the effect of hydrostatic pressure, the liquid medium surrounding cells or tissues serves to transmit the pressure equally in all directions. In contrast to this, mechanical pressure has a definite direction and its imposition on the cell is more or less accompanied with distortion of the protoplast. Mostly the effect of the mechanical pressure is to be understood as being exercised through deformation rather than through the direct effect of the pressure.

The older literature in this field is fully surveyed by LAUTERBACH (1921). LAUTERBACH (1921) noted in his own experiments using cells of *Chara* and *Nitella* that the mechanical impacts applied with unequal speed bring forth different effects on protoplasmic streaming even though the strength of the impact is the same. When the application of the pressure is sudden, the streaming stops, whereas when it is gradual, the protoplasm keeps streaming. The response of the cell to the mechanical impact is therefore to be distinguished depending on whether it is applied gradually or suddenly. The latter case is sometimes referred to as a mechanical shock and is usually accompanied by an action current (OSTERHOUT and HILL 1931). The visible effect of mechanical impact is then mostly a matter of the effect of an action current (see Section 11 in this chapter). LAUTERBACH's observations are readily understandable as being due to the action current elicited by mechanical pressure.

When pressure is applied gradually to a *Nitella* or *Chara* cell, the streaming usually does not stop even though the cell is deformed considerably, provided the cortical layer is not injured in the course of this procedure. LINSBAUER (1929) pinched an internodal cell of *Nitella* laterally by means of a "lateral compressor." The protoplasmic flow continues even though the cell walls on the opposite sides are gradually brought closer to each other so that eventually the protoplasmic layers lining them come in contact and fuse with each other. When the cell is completely pinched off, the rotational circuit is divided into two independent smaller circuits, but the speed of the streaming is the same as before. A similar experiment was also conducted by IWANAMI (1952, 1956) using pollen tubes of *Lilium auratum.* Such a pinching-off operation by means of a lateral compressor is the same in principle as a tying-off operation with a strip of thread, as has been described before (p. 27).

The effect of mechanical pressure on the plasmodium is very instructive in the investigation of the protoplasmic streaming in it. A simple and effective method to observe its response to mechanical pressure is to invert a cover-glass culture of a plasmodium (*Physarum polycephalum*) on a glass slide with enough water to keep the protoplasm from touching the slide and to push down gently at a certain point on the cover slip by means of a bent needle affixed to a micromanipulator (KAMIYA 1942). When some part of a plasmodium thus mounted first touches the surface of the slide, the movement of the endoplasm is strikingly affected. As the mechanical pressure is exerted unequally upon the plasmodium because of its uneven surface, the endoplasm is pressed out of the parts to which the pressure is applied and goes to those parts which are free from pressure. By this simple procedure the fluid endoplasm moves very readily and sensitively in the network of strands or in the branched courses throughout the fan-like expanses of gelatinous protoplasm. The artificial and passive movement of the protoplasm which has been induced mechanically is superimposed on the normal flow. Thus a temporary acceleration, retardation or reversal in direction takes place, after which the protoplasm takes its normal course.

Such a superimposed modified flow, controlled by mechanical pressure, causes no injury to the protoplasm. When the mechanical pressure is removed, the surface layer of the protoplasm takes on its original form again, presumably because of its elasticity, and the interior fluid protoplasm, which has been displaced, returns. By careful control of the screw of the micromanipulator, a spasmodic movement, as well as a rebounding of the interior protoplasm, can be repeatedly produced without giving rise to any observable disturbances, provided that the applied pressure is not so strong as to cause a structural disturbance in the protoplasm, such as was observed by BALBACH (1936) or by WINER and MOORE (1941). The latter authors found that pressure up to 166 gm./cm^2 causes reversible gelation of the flowing material. If the needle pushing down the cover slip is vibrated electromagnetically 120 times per second, the endoplasm is also forced to move back and forth presumably with the same period with an amplitude as large as 100 μ, and yet no injurious effect is observable afterwards.

From the simple experiment described above one can say that the streaming of the fluid endoplasm is modified in a manner directly dependent upon the pressure applied. These observations verify the conclusion of HILTON (1908), who applied mechanical pressure by the tapping of a needle upon a small piece of cover glass placed outside of the microscopic field and covering a part of the body of a plasmodium. The application of localized pressure can also be made with a blunt glass microneedle by pressing it directly against the surface of the plasmodium, as reported by

CAMP (1937). The evidence that a slight, uneven mechanical pressure induces artificial flow without any harmful effect, led KAMIYA to perform further experiments, in which air-pressure was used instead of mechanical pressure, and to construct the double-chamber already described (p. 40).

The effect of mechanical shock was also studied in the plasmodium of *Physarum polycephalum.* WINER and MOORE (1941) reported that sudden application of pressure or sudden release from slowly applied pressure results in gelation of the flowing material near the point of application. SEIFRIZ and EPSTEIN (1941) allowed a drop of water of known mass (36 mg.) to fall from a known height onto the plasmodium. When the slime mould is struck by a falling drop of water with sufficient momentum, all the protoplasmic streaming at and near the point of impact stops, although the protoplasm at a distance from the point of attack is kept flowing normally. The duration of standstill increases with the momentum of the falling drop, and is also dependent on the physiological state of the plasmodium such as vigor, age and development. A drop falling from a height of 15 cm. causes shock stoppage for 1¼ minutes in thin peripheral protoplasm. The shock effect is accompanied by sudden gelatinization of the protoplasm.

8. Gravity

When a cell of *Nitella* or *Chara* is placed with its long axis vertical on the stage of a horizontal microscope, we find a small difference in rate between the upward and downward streamings. This fact was noticed by EWART (1903) in several elongated plant cells, *Nitella, Elodea* and *Vallisneria.* In each case the upward stream is slightly, but definitely slower than the downward one. To quote an example of his data on *Nitella*: 32.6 μ/sec. (upward), 35.8 μ/sec. (downward), 34.5 μ/sec. (horizontal). In nearly all cases it was found that the difference between the horizontal velocity and the upward velocity is greater than the difference between the horizontal velocity and the downward velocity. In other words, the mean velocity of the ascending and descending streams is slightly less than that when the cell is horizontal.

BOTTELIER (1934) studied the effect of gravity on the protoplasmic streaming in *Avena* and confirmed EWART's observation. In *Avena* too retardation of the upward flow is slightly larger in amount than the acceleration of the downward flow. There has been as yet no satisfactory explanation accounting for this phenomenon.

EWART is of the opinion that the apparent difference in velocity observed in the ascending and descending streams has a purely physical origin, *i. e.,* they are due to the particles observed being different in density from the medium (endoplasm) in which they are suspended, heavy particles ascending slightly less rapidly than the upward stream, but descending more rapidly in the downward stream. With particles of lesser density, oil globules, etc., says EWART, the reverse is the case. In other words, EWART seems to believe that the ascending and descending streams of ground plasm, in which small particles and globules are suspended, move with the same rate no matter whether the stream goes upward or downward. This is, however, not likely to be actually the case.

Recently HAYASHI (1957) repeated some experiments along this line in the internodal cell of *Chara Braunii*. He confirmed EWART's observation and found a small but clear difference in the rate of particles between an ascending stream and a descending one in the cell placed with its long axis vertical. One of his data shows 95.4 μ/sec. and 98.9 μ/sec. for the ascending and descending streams, but he pointed out that the minute granules which are located in the direct proximity of the cortical gel layer of endoplasm move with an equal rate independently of the streaming direction, be it upward or downward, the fact showing a distinct difference in nature between the thin layer adjacent to the cortical gel and the remaining endoplasmic layer.

9. Centrifugal force

Investigations dealing with the effect of centrifugal force on protoplasmic streaming were made in *Elodea*, *Nitella*, etc.

In *Elodea* cells high centrifugal force brings about a stratification of the cell contents from the centrifugal end to the centripetal end in the following sequence: chloroplasts, hyaloplasm, nucleus and vacuole. Under the centrifuge-microscope the events occurring in the cell under centrifugal acceleration can be observed continuously. When a cell in active rotation is centrifuged, the protoplasmic stream toward the centripetal pole is retarded and the one toward the centrifugal pole is accelerated. When chloroplasts are evenly located on the 'roof' and 'floor' of the cell (epistroph) before the cell is centrifuged, they usually maintain their position in the cell under centrifugal acceleration up to about 500 times gravity or even more. Beyond the critical acceleration they start to fall rather abruptly followed by a rapid migration toward the centrifugal end. Such a behaviour allows us to assume that chloroplasts in the epistroph condition are anchored on the cortical gel layer in one way or another. When the cell is thrown into active rotation the chloroplasts must be deprived of this supposed adhesion or "anchor." There is a critical or "yield" acceleration which is around 500–700 $\times$ g for the chloroplasts in the epistroph state. In any event, the behaviour of chloroplasts under the effect of centrifugal force in this state is not like that of a particle freely suspended in a viscous medium.

The tolerance of the cell to high centrifugal acceleration is sometimes surprising. Using a ultracentrifuge, BEAMS (1949) ascertained that most *Elodea* cells survive exposure to as high as 350,000 times gravity for 30 minutes. In a cell examined immediately after centrifuging, Brownian movement is often clearly seen, but there is no sign of organized streaming. The inhibition of streaming, BEAMS supposes, may be due to a lowering of viscosity. The new streaming of the granular cytoplasm collected at one end of the cell is usually of an unorganized type; organized streaming is gradually established, which brings forth a slow irregular rotation of the large massed chloroplasts. The bunched chloroplasts separate later, and the streaming fully recovers to the normal state. It was also pointed out that in most of the cells the direction of streaming reoccurring after centrifugation remains the same. It is only in about 7% of the total cells that the direction is reversed after centrifugation (BEAMS 1949).

It is VIRGIN (1949) who first noticed in *Elodea* cells under a centrifuge-microscope that the return movement of the cytoplasm and chloroplasts starts before the end of centrifugation. It was then possible for him to obtain a rate at which the protoplasmic streaming—visible as a passive movement of the chloroplasts—can be made to stagnate. VIRGIN found that the centrifugal acceleration necessary to counteract the protoplasmic streaming is dependent on the consistency of the cytoplasm at the time of centrifugation. He used *Elodea* cells having two different relative viscosities of cytoplasm. A high viscosity value for the cytoplasm can be obtained by exposing the material to weak illumination (700 lux) for some hours. A low viscosity value is to be had by illuminating the cell with strong light (over 3200 lux) before centrifugation (VIRGIN 1948). As a measure indicating the relative values of viscosity, he adopted the time of centrifugation needed for the cytoplasm and chloroplasts to accumulate at the centrifugal end of the cells. The two groups of cells used for the experiments had relative viscosity values of 2 and 15. The mean speeds of centrifugation when the movement of protoplasm is entirely inhibited are 368 times and 200 times gravity for the cells having relative viscosities of 2 and 15, respectively. It was shown that the force necessary to counteract the protoplasmic flow appears to be greater for the cell whose protoplasmic viscosity is lower than for the cell with a relatively high viscosity. Thus he came to the conclusion that the motive force is inversely proportional to the viscosity of the protoplasm. This conclusion is based on the implicit assumption that the densities of the chloroplasts and protoplasm remain the same in the two groups of cells having different viscosities.

In another experiment mentioned in his paper VIRGIN reported that the rate of the return movement of the protoplasm is a function of the length of the cell and that the relationship between the rate of flow and the length of the cell is in direct proportion. This is shown by the fact that the distances covered by the chloroplasts plotted against cell length falls on a straight line. If the rate of streaming were the same for all the cells with the same viscosity value, the lines obtained would be parallel to the ordinate, *i. e.*, the distance covered by the protoplasm would be the same in all cells. VIRGIN believes that the slope of the straight line representing the relation of the cell length to the distance covered by the chloroplasts is a direct measure of the rate of the protoplasmic flow when actually the slope referred to has quite a different nature from the rate of movement. He represents graphically what he thinks to be the relation between the rate of movement and time (Fig. 2 in his paper). According to the graph, the return movement in a cell of low viscosity is more and more accelerated so that the rate increases to 10 times the initial value in 1¾ minutes.

Observation conducted by ourselves on *Elodea* cells under the centrifuge-microscope could not confirm this evidence.

The other material in which the effect of centrifugation has been well studied with special reference to the protoplasmic streaming is the cell of *Characeae.* It was LINSBAUER (1929) who first conducted experiments along this line. After subjecting the cell of *Chara foetida* to centrifugation for a short period of time, LINSBAUER observed that hundreds of endoplasmic balls are formed. The endoplasm which is pushed down to the centrifugal end of the cell comes back soon to its original position by way of reverse streaming. If, however, the cell is mechanically injured or its streaming is brought to a standstill with a shock just before centrifugation, the back streaming of the endoplasm is inhibited and the cell dies sooner or later. HAYASHI (1952) found that the normal state of streaming is completely restored only a few minutes after centrifugation in small internodal cells as well as leaf cells of *Chara,* but he reports that this is not generally true with an adult large internodal cell in which the streaming often stops after having covered a distance of a few millimeters, as coagulation of the protoplasm takes place. If only the large internodal cell, which otherwise cannot survive, is tied off beforehand into small fragments not exceeding 5 mm., each of these fragments can survive the centrifugation. Ligation can rescue the cell even though it is conducted after the cell is centrifuged, but only if it is done soon after the treatment near the centrifugal end of the cell where the endoplasm is accumulating. What can survive in this case, however, is the smaller fragment at the centrifugal end, which is rich in endoplasm, while the larger fragment at the centripetal end, in which there is only a trace of endoplasm, soon dies in spite of the intactness of the cortical layer.

Under the centrifuge-microscope BRECKHEIMER-BEYRICH (1949) successively observed in the leaf cell of *Nitella flexilis* the state of accumulation of protoplasm at the centrifugal pole. Just as VIRGIN (1949) observed in *Elodea,* BRECKHEIMER-BEYRICH (1949) reported that the endoplasm which has accumulated at the centrifugal end of the *Nitella* cell starts to move centripetally when the centrifugal acceleration becomes low. An acceleration of $30 \times g$ is just sufficient to prevent the endoplasm from flowing toward the centripetal direction. This critical acceleration is still lower for the *Nitella* cell in winter. HAYASHI (1957), who observed the internodal cell of *Chara Braunii* under a centrifuge-microscope of his own making, found that $40 \times g$ is enough to stop the centripetal streaming of most of the endoplasm, but that the minute particles which were streaming very near the cortical gel still flow in the centripetal direction under the same acceleration. This flow balances usually with $200 \times g$, but sometimes it continues to stream even under as high an acceleration as $600–800 \times g$.

As mentioned before (p. 39), this balance-acceleration can be widely changed according to the thickness of the endoplasmic layer as well as the difference in density between endoplasm and cell sap. Another important factor is the velocity distribution in the endoplasmic layer while the cell is being centrifuged.

Kamiya and Kuroda (1958 b) observed under the centrifuge-microscope that when the *Nitella* cell, either rhizoid or internode, is being centrifuged, the endoplasm frequently glides down over the cortical gel layer as a whole without causing any appreciable shear within the endoplasmic layer. What is inferred from this fact is that the interface where shifting takes place in the normal protoplasmic rotation is also the site at which shifting can occur most readily by external force (centrifugal force). On the basis of this fact Kamiya and Kuroda assumed that the viscosity of the boundary layer between cortical gel and endoplasm is lower than that of the endoplasmic sol itself.

Breckheimer-Beyrich (1949) noticed that the protoplasm collected at the centrifugal end of the cell into two layers, "A" and "B"; "A" is opaque-glassy and "B" is foamy. The "B" layer contains also protoplasm balls "C" which have been suspended in the middle part of the vacuole before the cell was centrifuged. In a series of sketches she described and showed the back migration of various parts of protoplasm occurring when the speed of centrifugation is made low or centrifugation is stopped. As for the plastids, which are embedded in the cortical gel, they do not come off even under 625 times gravity. Under the same centrifugal acceleration, the endoplasm stratifies to form "A" and "B" zones.

In her later paper, Breckheimer-Beyrich (1954) reported that the behaviour of the *Nitella* cell under centrifugation is not the same in summer and winter materials. In winter the "A"-plasm does not fall into balls under centrifuge treatment. Repeated centrifugation decreases the duration required for the endoplasm to be collected at the centrifugal pole and increases the speed of the return movement. It is important, as she herself stresses, that the return movement always occurs in the same direction as that before centrifugation. This fact forms the basis for her claim that the cortical gel plasm is responsible for the protoplasmic streaming in determining its direction due to possible regular orientation in the structure of its inner surface. We shall come again to this problem in a later chapter (p. 141).

Centrifugation can be applied also to an entirely different type of cell. Legler and Schindler (1940) have shown that in living diatom cells cytoplasm and plastids can be accumulated at one end of the cell through centrifugation. Cells treated in this manner can continue the sliding movement, but the direction of movement is always the same as that of the centrifugal treatment. This is because protoplasm can stream back along the central path (raphe) from the centrifugal pole in the body of the diatom with the result that the cell moves forward in the direction of centrifugation. This is an instructive experiment which contributes to elucidating the mechanism of locomotion of *Pennatae* diatoms, a rather special problem related to protoplasmic streaming.

Centrifuging is sometimes useful as a means of studying protoplasmic movement, because it forces to one end of the cell certain cell contents, which are in the way for the observation of protoplasmic motion. Küster (1933 a) points out that the centrifuged *Spirogyra* cell is a favourable material for studying protoplasmic motion. In the transparent part of such a cell, where only the vacuole and the cytoplasmic layer adhering to the cell wall are left, the cytoplasm forms a few clumps projecting into the vacuole. These clumps of protoplasm are the sites of not only a lively turbulent motion but also of a characteristic amoeboid motion.

10. Electric current

Passage of electric current through a cell brings about various morphological and physico-chemical changes in it. It has been known since the last century that a very weak current is sufficient to bring the protoplasmic flow in a cell of the *Characeae* to a standstill (cf. Velten 1876 c, Hörmann 1898). The effect is very different according to the intensity, duration and mode of application of the current. Perhaps it is pertinent to distinguish two cases, *i. e.*, the primary effect of electric current applied from the outside, and the effect of the action current elicited by electric shock. The latter case will be considered in the following section.

A constant current brings forth a marked effect on the cell, but the analysis of the effect is not simple. The amount of ionic current which actually passes through the cell mounted in an external medium depends in the main upon the electric resistance of the protoplasmic membrane, the reciprocal of which is the measure of ionic permeability. The resistance of the protoplasmic membrane being generally high in healthy cells, the current density in a cell in an electric field is usually only a small fraction of that in the external medium. Electric current (D. C.) causes pronounced polar changes in the distribution of anions and cations in the cell due to the existence of the semipermeable membrane. This in turn causes pronounced modification of the hydration-dehydration balance of the cytoplasm. Generally, the cytoplasm is hydrated at the cathodal end of the cell (*i. e.*, anodal side of the semi-permeable membrane), and dehydrated at the anodal side of the cell (*i. e.*, cathodal side of the semi-permeable membrane), with the result that the cell contents tend to be shifted to the anodal direction. If the current is switched off, the initial state is resumed quickly. In plant cells, the phenomenon is to be well observed in plasm-rich cells, especially in pollen-mother cells or in young staminal hair cells of *Tradescantia* (Kamiya 1937). This type of dislocation of the cell contents, which, though very conspicuous, is likely to have little to do with electrophoresis of the cytoplasmic particles, seems to be related with polar hydration and dehydration of the ground plasm.

The effect of D. C. was studied on the *Elodea* leaf by two different groups of authors, Heilbrunn and Daugherty (1939) and Tobias and Solomon (1950), with rather different results. Heilbrunn and Daugherty reported that chloroplasts migrate either to the anode or the cathode and accounted for the variability as being due to a photosynthesis effect. With strong currents, the rotational streaming comes to a standstill; with weaker current, the rate of streaming is affected. Heilbrunn and Daugherty noticed some differences in the rate of movement of chloroplasts on the opposite wall of the cell according as whether they move toward the anode or the cathode when a weak D. C. field (0.1–0.7 mA) is applied to the leaf with its longitudinal axis parallel to it. Under conditions inhibiting or retarding photosynthesis, chloroplasts move invariably more rapidly toward the cathode, but the sign of the difference is not always uniform when the cell is in the light. Thus they came to the conclusion that the chloroplasts are charged in the presence of light either positively or negatively, but that they become negatively charged if the pH of the cell gets higher with reduction in the amount of carbon dioxide in the cell as a result of photosynthesis.

The work of Tobias and Solomon (1950) is mainly concerned with the polar change in viscosity under D. C. field. They noticed that the D. C. (max. 110 V) affects the movement of chloroplasts, cytoplasm, vacuole and vacuolar inclusions differently. According to them, an early stage of observable change is cessation of the motion of chloroplasts carried by protoplasmic streaming with subsequent migration toward the anodal end of the cell. In contrast to the observations of Heilbrunn and Daugherty (1939), these authors state: "Whenever we have seen chloroplast displacement (in over 225 cells) it has been toward the anode. Plastid standstill and agglomeration may be complete while the protoplasm continues to stream in an apparently normal fashion. When the protoplasm itself stops, it clearly slows down and stops first at the anodal end of the cell ... As the plastids move anodally the vacuole moves cathodally and the cytoplasm piles up anodally. In our experiments it seems clear that the plastids have themselves been negatively charged."

As pointed out by Heilbrunn (1952) the current applied by Tobias and Solomon was probably much stronger than that used by Heilbrunn and Daugherty. The streaming, however, still continued and the effect was reversible under the conditions set up by Tobias and Solomon.

The effect of electric current was studied also in myxomycete plasmodia with special reference to the streaming. Using the double-chamber, Kamiya and Abe (1950) induced electric potential difference between the two protoplasmic blobs of a plasmodium in the two compartments, and measured how the motive force is modified by the electric current applied. The DPG pattern, however, is not influenced appreciably when the potential difference up to 4 V or even higher is applied for a short period of time. This is 1000 times as high as the electromotive force which the plasmodium produces spontaneously by itself. Yet the motive force responsible for protoplasmic streaming is not changed. Obviously, the potential difference applied externally between the anterior and posterior parts of the plasmodium has nothing to do with the cause of the protoplasmic streaming (cf. p. 159). In other words, protoplasmic streaming in the plasmodium is not explained in terms of the electroendosmotic phenomenon.

If, however, a weak current is applied continuously for a long period of time, the current is not without effect. Watanabe, Kodati and Kinoshita (1938) studied the effect of weak direct current on plasmodia of eight different kinds of *Myxomycetes* and found that they all tend to move toward the cathode (negative galvanotaxis). ANDERSON (1951) also studied the galvanotactic behaviour of plasmodia using *Physarum polycephalum*. Orientation of the mould is always cathodal in the D. C. field in the range of 1.0 to 8.0 μA/mm^2. Later ANDERSON *et al.* (1953) analysed the cation contents of the mould after it had migrated for approximately 3 hours. What they found is the rather unexpected fact that calcium concentrations in the anterior portions are significantly higher both in transplants in the electric field and in the controls. Sodium was found in higher concentrations at the posterior regions, which are anodal in the transplants in the electric field, both in the experimental mould transplants, and in the controls. Potassium concentrations at the cathodal (anterior) regions, however, are significantly higher than at the anodal region in the experimental transplants, while no significant difference in potassium concentration was found between the posterior and anterior regions in the controls. ANDERSON's experiment thus shows the remarkable fact that the K^+ is readily transported within the plasmodium placed in the electric field while Ca^{++} and Na^+ are comparatively stable.

SEIFRIZ (1939 a) observed a remarkable orientation of protoplasmic particles into parallel rows in a myxomycete plasmodium placed in an electric field.

11. Action current

The action current being of cell origin, it might be inadequate to deal with the relation of it to the protoplasmic streaming in this Chapter entitled "Influences of External Factors." Usually, however, it is generated, if not always, by external stimuli, and for the sake of convenience the matter is discussed here.

It is HÖRMANN (1898) who first discovered that the action current in the *Nitella* cell is accompanied with an important change in the mobility of the protoplasm. Various aspects of this remarkable phenomenon were studied more exactly later by AUGER (1931, 1939), UMRATH (1933, 1953), HILL (1941), SIBAOKA and ODA (1956), KISHIMOTO and AKABORI (1957), etc.

Results showing close relationship between the action potential and cessation of the streaming have been reported in various species of *Characeae, e. g., Nitella flexilis* (HARVEY 1942, KISHIMOTO and AKABOLI 1957), *N. mucronata* (UMRATH 1933), *N. opaca* (UMRATH 1953), *N. translucens* (AUGER 1939), *N. stelligera* (AUGER 1939) and *Chara Braunii* (SIBAOKA and ODA 1956).

Generally, the streaming stops abruptly when an action current appears but the stoppage takes place somewhat later than the beginning of the action current. According to FRANCK and AUGER (1932) who studied the phenomenon using a cinephotomicrographic method, the beginning of the action potential precedes the abrupt change in the speed of the protoplasmic streaming by 1.2 to 2 seconds. A simultaneous record of the action potential and the time of cessation of the streaming which was obtained recently by SIBAOKA and ODA (1956), shows a similar result, namely, the cessation occurs just when the initial rising phase of the action potential lasting about 1 second is over, or, at the point almost coincident with the crest of the potential curve.

After the abrupt change of the speed, which occurs in less than $^1/_8$ second (Frank and Auger 1932), the streaming starts again; in *Nitella stelligera*, for instance, the full recovery of the protoplasmic movement takes at least 5 to 10 minutes. The time course of the recovery process in the rate of flow may well be different according both to the material and external conditions. Auger (1939) described two different recovery courses. Fig. 61, which was worked out by Sibaoka and Oda (1956), shows how the streaming in *Chara Braunii* stops and recovers after induction shocks of unequal strengths were applied. Streaming stops suddenly with the generation of

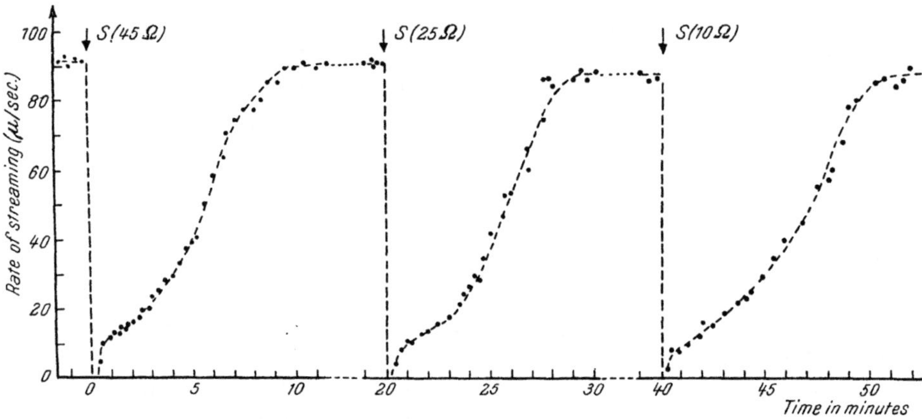

Fig. 61. Abrupt cessations and gradual recoveries of streaming with action potential through induction shocks of various strengths. Application of shock indicated by arrows. Shock strengths shown in Ω. Threshold value of this ce is 50 Ω. Temp. 25°C.
(Sibaoka and Oda 1956.)

action potential. Protoplasm remains at a standstill for 15–20 seconds before it starts moving. As is shown in Fig. 61, the streaming, when it starts anew, is slow at first and is gradually accelerated till full recovery is attained after about 10 minutes independently of the strength of the induction shock. Sometimes, however, the full recovery of the rate of protoplasmic movement occurs so quickly that it is accomplished within several seconds, almost coinciding with the recovery of the potential (Auger 1939). In an autumn material Sibaoka and Oda (1956) themselves also frequently observed a case where the streaming does not stop completely but slows down abruptly and returns quickly to the original rate. Also in this case, the form of the recovery course has nothing to do with the strength of the stimulus. In other words, the stop or slow-down response of the streaming obeys the all-or-none law.

To this, however, we have some exceptions. Umrath (1953) found that in young internodal cells of *Nitella opaca* the intensity of action potential increases stepwise with the strength of stimulus. In such a case the duration of the temporary cessation of streaming is also dependent on the intensity of action potential varying mostly from 50 to 150 seconds. Thus we see here that not only the action potential but also the stop response of the streaming and the time required for the recovery of the streaming are all dependent on the strength of the stimulus applied.

The experiments of Sibaoka and Oda (1956) and of Kishimoto and Akabori (1957) show that protoplasmic streaming in the cell of *Chara* stops temporarily only when and only where the action potential is elicited. It is clear in the light of their experiments that the temporary stoppage of flow is not a direct result of the stimulus applied but of some yet unknown protoplasmic changes occurring as a consequence of the action potential. As stated above, generation of an action potential is followed invariably with the abrupt stop or retardation of the protoplasmic streaming, but the reverse is not always true. When the electrical current applied for stimulation is small enough, the streaming in *Nitella* slows down and gradually comes to a stop without eliciting any action potential (Hill 1941). It is also to be stressed that electrical response may be obtained before protoplasmic streaming regains its normal rate. This important fact has been pointed out first by Auger and further verified by Umrath (1933), Hill (1941), Kishimoto and Akabori (1957), etc. Under certain conditions, on the other hand, *Nitella* cells lose their ability to give an electrical response to stimulation—for instance, when the cell is placed in distilled water for days (Osterhout and Hill 1933),—yet the protoplasmic streaming continues.

From all these facts we may conclude that, first, absence of streaming is not a sign of failure of electrical response, and second, presence of streaming does not necessarily indicate electracal irritability. What we may say is that there is no action potential without sudden stop or retardation of the streaming.

In seeking the cause of the sudden stop of flow, we may think of two possibilities, the spontaneous gelation of the endoplasm involved in the streaming or the temporary upset of the mechanism responsible for the flow. In this connection a fact worth noting is that reported by Umrath (1933) that during this temporary cessation of flow the viscosity of the endoplasm is not changed. This conclusion is based upon an analysis of Brownian motion. The cessation of flow must therefore be accounted for not by the sudden setting of the protoplasm but by the temporary impairment of the motor mechanism of the streaming. The problem as to why generation of action potential impairs the driving mechanism is still open to discussion. Temporary ionic loss or shift accompanying the action current may have a significant bearing upon this problem, but no experimental facts are available at present to throw further light on it. It may be, as suggested by not a few workers, that the cessation of flow provoked by the action potential is comparable to the contraction of muscle fiber. If so, the problem in question may have some relation to the interlinking mechanism between excitation and contraction of a muscle fiber.

In addition to the studies on *Characeae*, the relation between the electric response and the protoplasmic streaming has also been studied in the myxomycete plasmodium. Tasaki and Kamiya (1950) found that the electrical response released by the plasmodium of *Physarum polycephalum* through mechanical and electrical stimuli disobeys the all-or-none law and is always accompanied with a simultaneous change in the electric impedance of the protoplasm, similarly to the electrical response in *Nitella* (COLE and CURTIS 1938) and other cells. There is, however, no sign of transmission of response along a strand of the slime mould. Protoplasmic streaming proceeds unhindered at the moment when electrical responses are induced. Further there is no indication of microscopically observable changes in protoplasm at the moment of response. This is probably due to the fact that the local stimulation of the submerged *Nitella* cell can be rapidly propagated to the entire cell while in the myxomycete plasmodium the region of response is limited mostly to that locus where the protoplasm is directly stimulated from the outside. Even though the "action potential" exerted a profound effect on the protoplasmic flow, the intracapillary flow of the endoplasm in the plasmodium would be under the control of a region which does not take part in the electrical response.

As a matter of fact, DIANNELIDIS and UMRATH (1953), who studied the electric potential difference between the outside and inside of the plasmodium of *Physarum polycephalum* and its electrical response (cf. also TAUC 1953, 1954), observed that protoplasmic streaming comes to a standstill with a latent period of many seconds (as much as 20 sec.). This is probably due to the difference in conditions, for in the experimental setup of TASAKI and KAMIYA only a very small area (less than 0.5 mm^2) of the strand is thrown into action while in that of DIANNELIDIS and UMRATH the plasmodium was presumably excited over a wide area including the seat of the motive force. It is probably to be expected that the protoplasmic streaming is subject to a pronounced influence of the action potential if the seat of the motive force is within the action sphere of excitation.

12. Magnetic field

EWART (1903) could find no perceptible change in the rate of flow in *Vallisneria* and *Chara* cells when they were brought to a magnetic field, the long axes of the cells being parallel to the lines of force. But if the cells are arranged so that the direction of streaming cuts the lines of force at right angles, the rate of flow is affected to a considerable extent after 10–20 minutes. The ultimate effect is always retardation; frequently, the streaming in *Chara, Nitella, Elodea* and *Vallisneria* stops after very long exposures, if the protoplasmic stream cuts the lines of force at right angles. EWART interpreted this effect as being due to the weak electric current which may be produced by the passage of protoplasm across the lines of force. The fact that a greater relative retardation is brought about when streaming is rapid (*viz.* at 30° C.) than when it is slow (*viz.* at 10° C.) is explained in terms of increased current which is to be induced in proportion to the rate of protoplasmic flow across the magnetic field.

SSAWOSTIN (1930), who surveyed old magnetophysiological literature, conducted a series of experiments similar to EWART's on the effect of a magnetic field on protoplasmic rotation using *Vallisneria, Elodea* and

Chara. He also noted that the effect of the magnetic field on the rate of streaming differs according to the orientation of the cell in the field. If the protoplasm streams in a strong magnetic field (7,000 Gauss) parallel to the lines of force, the rate is always decreased to a considerable extent, usually by 15–30%, which was not noticed by EWART. When the cell takes such an orientation that the protoplasm streams at right angles to the lines of force, an appreciable effect is also brought about on the rate. But the effect does not show a constant tendency in this case; it is either accelerated or retarded with an equal probability. The absolute amount of acceleration is greater (sometimes as far as 50%) than that of retardation (usually less than 30%); the acceleration is likely to be less dependent on the strength of magnetic field than the retardation.

The effect caused by the magnetic field appears either immediately or after a certain lapse of time. In the latter case it is suggested that the effect is mediated with some chemical stimulation. On the other hand, SSAWOSTIN noted also that the protoplasm responds with its motion to the reversal of the direction of magnetic field. On the basis of this fact he concluded that the moving protoplasmic layer carries an electric charge. Thus, according to SSAWOSTIN, the magnetic field exerts its effects on the protoplasmic rotation through two different ways: namely, first, retardation is brought about through increase in friction between flowing and resting (cortical) protoplasmic layers caused by the force generated electromagnetically, and second, acceleration of flow is induced through a stimulation of anoxidative reaction between paramagnetic substances in the cell (Fe, Mn, and O_2). SSAWOSTIN explains that whether the streaming is accelerated or retarded is determined by whether "ponderomagnetic" effect or "biochemical" action is predominant.

The above interpretation of SSAWOSTIN is rather different from that of EWART. In spite of SSAWOSTIN's explanations, the question is still left open as to why the effect is sometimes acceleration and sometimes retardation when protoplasm streams at right angles to the lines of force; we do not yet know under what conditions the effect is acceleration only and under what conditions it is retardation only. Magnetophysiology has so far thrown little light on the understanding of protoplasmic streaming and it still remains an unexplored field of study.

13. Hydrogen ions

Not a few workers investigated the influence of pH on the rate of protoplasmic streaming, but the results are rather diverse partly due to the characteristics of the cell observed and partly to the kinds and concentrations of the buffer solutions used.

STRUGGER (1926) reported that hydrogen ion concentration exerts a marked effect on the protoplasm of the root hair of *Hordeum vulgare* even in a narrow zone around the neutral point (pH 6.80–pH 7.55). He used K-phosphate buffers of about $^1/_{15}$ M. The effect is observable in its visible colloidal states as well as the rate of the streaming. The minima of the streaming rate are found at pH ranges 6.85–6.90, 7.00–7.05 and 7.35 where the flocculation of plasmacolloid is found to be maximal. The maxima of the rate are found at pH 6.70–6.80, 7.15–7.20 and 7.40–7.45 where flocculation is minimal. This work was further extended by using Na-acetate buffer (STRUGGER 1928). Between pH 5.60 and pH 6.90 he established two maxima of the rate of flow at pH 5.85 and 6.40. At pH 5.10–5.20 and pH 5.50–5.65 the streaming stops. Change in the rate occurs concomitantly with morphological changes of the protoplasm, *e. g.*, change in the volume of protoplasm (hydration), the formation of new granules etc. By comparing different buffers (Na-acetate buffer and phosphate buffer) STRUGGER arrived at the opinion that hydrogen ion concentration is primarily responsible for these changes although the interfering effect of anions cannot be neglected.

DOI (1950) found in his studies on the protoplasmic streaming in root hairs of many crop plants that the number of root hairs of *Oryza* exhibiting protoplasmic streaming decreased below pH 4.0 and above pH 10.4, in well water. In his experiment the pH was adjusted with H_2SO_4 or KOH. The streaming stopped completely at pH 1.4 as the result of coagulation of the protoplasm. At pH 10.5 the protoplasm of all root hairs dissolved. The pH values of acid well water necessary to bring forth stoppage of the protoplasmic streaming were lowest with *Oryza,* highest with *Hordeum,* and intermediate with *Triticum.* His experiments show that, for root hairs of the crop plants resistant to soil acidity, relatively low pH values are necessary in order to arrest the streaming. The rate of streaming in root hairs of *Oryza, Triticum,* and *Hordeum,* contrary to the STRUGGER's result, changes only slightly with the external pH, between 4 and 9.

Results somewhat similar to those of STRUGGER were obtained by COLLA (1929) who studied the effect of hydrogen ion concentration on protoplasmic streaming in internodal cells of *Chara crinata* using Na-phosphate buffers. She found that there are two optima for the rate of streaming, *i. e.,* pH 5.3 and 7.1, a slight minimum occurring at pH 6.2, where the cytoplasm exhibited some flocculation. PFEIFFER (1931) reported that in *Nitella cernua* and *N. flexilis* the pH of the external buffer solution (MCILVAINE) in which the rate of streaming reaches the maximum is the same as that in which the viscosity of protoplasm is at the minimum. The pH, which is 5.2, is assumed to be the IEP of protoplasm.

ROBBINS (1926) studied the effect of pH on the rate of protoplasmic flow in *Elodea* leaves. Using thirteen sodium phosphate buffer solutions containing 0.03 M sodium in the pH range from 4.8 to 7.2, he found maximal rates at pH 4.8 and 5.6; the rate of flow decreases in the solutions above pH 6.0. On the basis of the data obtained using acid and basic dyes he supposed that the isoelectric point (IEP) was at pH 6.0–6.2, but he was unable to prove any differences in viscosity in relation to the pH.

Literature on the *Avena* coleoptile shows that the protoplasmic streaming in the *Avena* coleoptile is rather insensitive to the pH of the external solution. Neither acetic acid at concentrations between 1×10^{-3} and 1×10^{-6} M (THIMANN and SWEENEY 1937) nor the addition of alkali to give a pH of 9 (SWEENEY 1941) have any effect on the streaming. RIETSEMA (1950) studied the effect of pH on the protoplasmic streaming in the *Avena* coleoptile very exactly using 0.01 M phosphate buffer solution and proved that between pH 4 and 8 the rate of stréaming is independent of the external pH. KELSO and TURNER (1955) who worked with staminal hairs of the same plant found also no change in the rate of streaming when the cells are transferred to phosphate buffers at pH 4 and 7.

The response of the protoplasmic flow in the slime mould to the change in pH is also not very sensitive. LEPOW (1938) reported that the rate of streaming in *Physarum polycephalum* is not much altered between pH 5 and 9. Dynamoplasmometric analysis showed that the motive force responsible for the protoplasmic streaming in this organism does not change conspicuously following a change of pH between 5 and 8 (NAKAJIMA, unpublished).

As is the case in all other chemicals, an important problem in connection with the effect of pH is how far the pH of the cell inside (protoplasm and vacuole) is dependent on the pH of the outside. BECKER (1936) presented a table showing that the pH of the vacuole of the staminal hair of *Tradescantia virginica* is influenced by the pH of the external medium according to the difference in acids used. In carbonic acid, valerionic acid and acetic acid the streaming ceases even though the colour of the indicator (natural anthocyanin) corresponds to the normal state of the hairs. Obviously, these acids, especially the two organic acids, are highly toxic. As to the effect of carbonic acid, which is said to have an anaesthetic effect, more will be stated in the next section. Generally, even though the pH of the vacuole is changed to some extent under the influence of acids in the external medium, the pH of the cytoplasm may remain unchanged before death ensues. This is presumably due to its buffer action. At any rate this explains why no effect is observable in the streaming rate on addition of various acids including mineral acids and weak organic acids at a wide pH range in *Tradescantia*, *Avena*, etc.

In RIETSEMA's experiments on *Avena*, as stated above, the rate of streaming remained the same between pH 4 and 8 when $^1/_{100}$ M phosphate buffer was used, but it was inhibited by acetic acid. Toxicity of Na-acetate buffers was also shown by SAKAMURA and KANAMORI (1935) in root hairs of *Brassica chinensis*. They pointed out that when weak acid is administered to root hairs the effect of primary importance is that of undissociated acid molecules rather than that of hydrogen ion. Thus we see that the relation between the pH of the outer medium and the rate of flow is different according to the specificity of the cell and to the specific effect of the buffer. Roughly speaking, however, the streaming in many plant cells is greatest in weak acid solutions and decreases in neutral or alkaline solutions.

14. Carbon dioxide

A pronounced effect of carbon dioxide on protoplasmic streaming was noted by earlier authors. KLEMM (1895) observed a very rapid cessation of streaming when root hairs of *Trianea* or leaf cells of *Vallisneria* were suddenly subjected to CO_2 gas or to water saturated with CO_2. EWART (1903) noticed that a reversible shock-stoppage ensues upon sudden exposure to pure CO_2. In hairs of *Tradescantia, Cucurbita, Urtica* etc., it takes a few minutes before stoppage occurs owing to the presence of a cuticule which prevents diffusion (cf. LOPRIORE 1895), while in *Vallisneria, Elodea, Trianea* or pollen tubes, all of which have no cuticule on the cell surface, a still shorter period of time is needed before the streaming comes to a standstill in CO_2. Cells of *Chara* or *Nitella* are more resistant to the action of CO_2; the streaming is neither retarded after 5–15 minutes immersion in pure CO_2 nor is it completely stopped before 20–30 minutes (EWART 1903). The effect is usually perfectly reversible.

BECKER (1936) showed that the movement of protoplasm in the staminal hair of *Tradescantia* ceases in CO_2 gas when the colour of the indicator (natural cell sap) corresponds to the normal pH of the hair, *i. e.*, around pH 4.5. Thus it is clear that the characteristic action of CO_2 is not due to its pH.

The cessation of flow brought about by CO_2 is reversible if the material is not subjected to it for too long. When the streaming is once allowed to restart in the air, the new streaming does not stop even in a current of pure CO_2 for hours. Generally, cells treated with CO_2 tend soon to accomodate themselves to very high percentages of carbon dioxide.

SEIFRIZ (1941, 1942) paid special attention to the striking action of CO_2 gas on the myxomycete plasmodia. An important feature of the action is the abrupt cessation of streaming. When CO_2 gas, say 50% diluted with air, is passed through the observation chamber in which the slime mould is spreading, the streaming stops after 5–20 seconds all of a sudden with little forewarning. Although the abrupt cessation of flow occurs some time after CO_2 is applied, this is also a sort of shock stoppage. No abrupt stop occurs when the concentration of CO_2 is raised gradually. On removal of CO_2, normal activity is soon resumed with no indication of damage. SEIFRIZ (1941, 1942) regarded this state of "suspended animation" as anaesthesia.

It was found out with the aid of the double-chamber method that this cessation is nothing but the instantaneous gelation or setting of protoplasm, since pressure externally applied in that state fails to cause the endoplasm to stream passively, which is not the case when the endoplasm is in a state of sol. Furthermore, we have cogent facts showing that the mechanism generating the motive force is still functioning when protoplasm is in a state of stiff gel under the effect of CO_2 (KAMIYA and NAKAJIMA 1955) (cf. p. 59).

No gelation occurs if the concentration of CO_2 gas is low. It was clearly shown, however, through the measurement of the motive force that the slime mould reacts very sensitively to diluted carbon dioxide. The effect is most exactly represented by the method of dynamoplasmometry. The dynamoplasmogram shows that a pronounced change in polarity has taken place under the influence of CO_2. Referring to Fig. 62, we see that the undulating curve shifts as a whole to the upper side of the base-line as soon as 10% CO_2 gas is administered to one half (*a*, in Fig. 22) of the plasmodium. This means that the motive force is always exerted in one direction from the CO_2 side to the air side. In spite of such a situation the curve shows clear periodic variations during the treatment. The shift of the undulating curve to the upper side of the base-line is an indication of the negative taxis of the slime mould in respect to CO_2, because we need more positive counter-pressure on the air side (compartment *B* in Fig. 22) in order to prevent the streaming from the CO_2 side to the air side.

Simultaneous measurement of the potential difference between the two parts of the plasmodium reveals that the half of the plasmodium treated with CO_2 becomes always higher in electric potential than the other half which is left untreated. Therefore, we see in this case that the protoplasmic

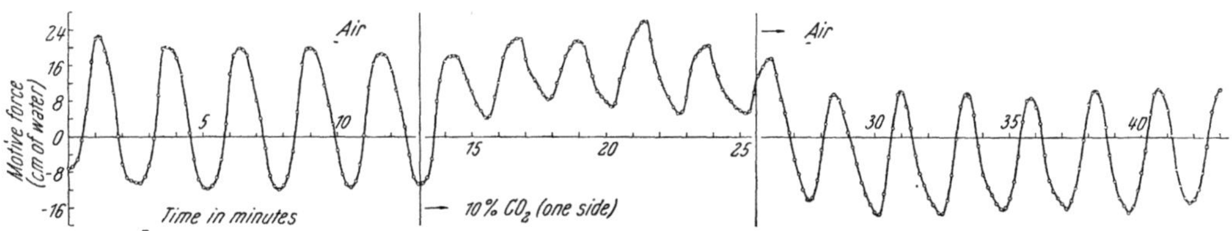

Fig. 62. Generation of the motive force when only one half of the plasmodium was treated with 10% CO_2. (KAMIYA 1953.)

flow tends to occur from the higher potential part to the lower potential part. This is a characteristic of the CO_2 effect inasmuch as in cases treated with most reagents other than CO_2 the opposite is true, namely, the polarity of the motive force is directed from the lower potential side to the higher potential side.

Another fact which interests us in respect to CO_2 action was shown by LOEWY (1950). As already touched upon (p. 66), he pointed out that the plasmodium of *Physarum polycephalum* disintegrates in pure nitrogen gas in less than 80 minutes, but if a small amout of CO_2 gas is added, the harmful effect of pure nitrogen is avoided, the plasmodium showing a vigorous streaming for hours without oxygen. Just how CO_2 removes the toxic effect, however, is still to be dealt with.

15. Salts

Among the old literature dealing with the effect of salts on the rate of streaming, COLLA's work (1929) is worthy of note. She conducted extensive studies on the effect of 0.1 M solutions of a variety of salts on the rate of protoplasmic streaming in *Chara crinita*. It was shown that chloride salts of Na, K and Mg all increase the rate of flow in the following order: K > Na > Mg. LiCl, $CaCl_2$ and $BaCl_2$ on the other hand, retard the flow and bring about a rather quick cessation. Barium chloride and lithium chloride are very toxic to young *Chara* cells. The effect of the anion series was studied using sodium salts. They all stimulate the flow, the stimulating effect being in the following order: $NO_3^- > (HPO_4^{--}, H_2PO_4^-) > SO_4^{--}$, $(HSO_4^-) > Cl^- > CO_3^{--}$, (HCO_3^-). According to GIMESI and POZSAR (1955), the rate of flow in *Elodea* cells is retarded under the effect of 0.02 M solutions of the following salts in varying degrees from 74% down to 62% of the control rate in water; the rates in μ/sec. are shown in parentheses: water (21.8), KNO_3 (16.1), KCl (15.9), K_2SO_4 (18.0), $Ca(NO_3)_2$ (13.3), $MgSO_4$ (13.8), $AlCl_3$ (13.3).

Generally, the analysis of the effects of different salts involves complications on account of various factors, such as unequal permeability of the cell to different salts or ions, possible differences in osmotic pressure or in pH of equimolar solutions of different salts, etc. Another complication in comparing the effect is that the time courses of the change in rate are different according to the salts used.

SEIFRIZ and URAGUCHI (1941) studied the toxic effect of heavy metals on the slime mould, *Physarum polycephalum* using their nitrate salts. They adopted the time required to stop protoplasmic streaming as a criterion of toxicity. The order of toxicity ascertained for the metal cations is: Ag > Hg > Cd > Tl > Cu > Pb > Zn > Y > Sr > La > Rb. Five of the above heavy metals are adsorbed by blood charcoal in the order: Ag > Hg > Cu > Pb > Zn. Comparison of the order of adsorption of these salts with the order of toxicity shows a direct relationship between adsorption and toxicity. URAGUCHI (1941) reported the formation of remarkable rhythmic bands in the plasmodium through the precipitation of protoplasmic proteins when the plasmodium is bathed in these salts or when they are injected.

It is also known that heavy metals, such as Cu or Sr, both of which are highly toxic, are able to initiate the protoplasmic streaming in *Elodea* or *Vallisneria* (SEIFRIZ 1922, FITTING 1925). The streaming induced by these salts, which is obviously pathological, goes on furiously for days until the cell dies.

The hydration phenomenon of protoplasm in plant cells caused by alkali salts has been especially well studied by HÖFLER (1928, 1934). When a small piece of a section of onion epidermis is brought in pure, hypertonic solution of KCl, KNO_3 or NaCl for many hours, there occurs often the phenomenon called "cap plasmolysis". If the solution is hypotonic, vacuole contraction is often brought about. The protoplasm is strongly hydrated under the influence of the invading alkali salts and enlarges its volume at the expense of the cell sap.

No streaming is to be observed in the strongly hydrated protoplasm. Nevertheless the hydrated protoplasm is still alive and can be brought back reversibly to the active state. As has been studied by HÖFLER (1939) and KAISERLEHNER (1939), this can be done by treating it with the solution of a calcium salt. In the cell, which is once transferred into the $CaCl_2$ solution, hydration goes back toward the normal state and the motility of protoplasm is resumed (HÖFLER 1940). This vital change of the colloidal state, in which protoplasm increases to 5–10 times its original volume under the effect of a potassium salt, is accompanied with a marked decrease in viscosity. Here again we see a good example showing that excessive hydration accompanied with abnormally low viscosity of protoplasm impairs the generation of the force producing the streaming.

The role of bivalent cations can be studied not only by applying them to the cell, but by depriving it of them through the application of chelating agents either from without or by means of microinjection. If we measure the motive force of flow in the plasmodium of *Physarum polycephalum* after injecting 10^{-3} M ethylenediaminetetraacetic acid (EDTA) of $1/10$ the volume of the plasmodium, we find that the motive force decreases conspicuously (TAKATA 1957). The motive force, which has been once diminished through the injection of 10^{-3} M EDTA, recovers almost instantly to the normal strength if a small quantity of $MgCl_2$ is further injected. Neither ATP nor Ca^{++} is effective for this recovery. It is likely from this fact that a small amount of magnesium is essential for the production of the motive force in the slime mould.

Recently OHTA (1958) studied the effect of EDTA and diethyldithiocarbamate (DDC) on the motive force by using the double-chamber method. He applied these chelating agents from the outside. They both diminished the motive force at the concentrations of 10^{-2}–10^{-3} M. Upon application of Mg^{++} from the outside after the removal of EDTA the diminished motive force not only recovers, but rather increases as compared with its original state. The recovery by Mg^{++} is not seen in the plasmodium treated with DDC.

On the other hand, according to TAKATA (1957), the motive force increases rather than decreases if more diluted EDTA (5×10^{-4} M) is injected. The amplitude thus augmented through the injection of dilute EDTA, however, is brought back to the normal level by applying Ca^{++}. Thus Ca^{++} inhibits the increase of the force brought about through dilute EDTA. This effect is not replaced with Mg^{++}. As it is known that EDTA has a greater affinity for Ca^{++} than Mg^{++} (BOZLER 1954), we can reasonably assume that the major cation which is chelated with dilute EDTA in the plasmodium is calcium and a greater part of magnesium remains still free or unchelated. If we make a further assumption that Ca^{++} acts antagonistically against Mg^{++} in producing the motive force, we can well understand that the motive force that has been once augmented through diluted EDTA is again suppressed to the normal level through the application of Ca^{++}. Studies dealing with the problems of salts, especially the effects of Ca^{++} and Mg^{++} on the production of the motive force are very important as they have a close relation to the mechanochemical system (p. 160) of the protoplasm and its ATPase activity.

16. Auxins

Many authors have found that various auxins increase the rate of protoplasmic streaming in the epidermal cells of *Avena* coleoptile and other plant cells at low concentrations.

THIMANN and SWEENEY (1937 a, b) reported a marked effect of indole-3-

acetic acid (IAA) on the rate of streaming when present in concentrations between 0.5 and 0.002 mg./l., maximum increase being brought about at 0.01 mg./l. Higher concentrations of IAA retard or stop the streaming. This work has been further extended by SWEENEY and THIMANN (1938, 1942) and SWEENEY (1941, 1944). OLSON and DUBUY (1940), however, failed to obtain stimulation of streaming with IAA, although they observed slight inhibition with higher concentrations. SWEENEY (1941) ascribed the negative results of OLSON and DUBUY to their use of coleoptile sections infiltrated with water, as she found no acceleration of streaming by IAA if the intercellular air space of the coleoptile is replaced with water [7].

KELSO and TURNER (1955) studied this problem further in the mature staminal hair-cells of *Tradescantia* and found a striking parallelism with the results obtained by THIMANN and SWEENEY with the *Avena* coleoptile. The main quantitative difference lies in the concentrations of IAA necessary to cause similar effects in the two tissues, namely, for *Tradescantia*, the maximum effect of the auxin is brought about at a concentration of 1 mg./l., whereas for *Avena* the maximum effect is produced at 0.01 mg./l.

The optimum concentration of IAA for the acceleration of protoplasmic streaming is very different according to the material. In root hairs of *Avena* it is from 10^{-4} to 10^{-5} mg./l. (SWEENEY 1944). This is considerably lower than the optimal concentration for the streaming in coleoptile cells, but is of the same order as that which is said to bring about increase in the rate of elongation of roots (AMLONG 1936, FIEDLER 1936, THIMANN and LANE 1938). Generally speaking, there is a parallelism between the effects of auxin on growth of roots and its effects on streaming in root hairs in that IAA concentrations causing optimal acceleration for these structures are much lower than those effective for streaming and for growth in coleoptiles. Also the lowest concentration bringing about inhibition is smaller for root hairs or roots than for coleoptiles, both in so far as streaming and growth are concerned (SWEENEY 1944).

According to the recent work of SANDAN and OGURA (1957) the maximum rate of flow in *Nitella* is obtained at the concentration of 0.1–0.2 mg./l., which range is also optimal for the growth of both newly formed shoots and rhizoids. For comparison the relations between the concentrations of IAA and its effects on both protoplasmic streaming and growth in *Avena* and *Nitella* and streaming in *Tradescantia* hair are given in Fig. 63. Except the data on *Nitella* (SANDAN and OGURA 1957) which are added, this figure represents the generalized scheme elaborated by KELSO and TURNER (1955) based on the data from THIMANN and SWEENEY (1937), SWEENEY (1944), AUDUS (1953) and KELSO and TURNER (1955) themselves.

The effect of IAA was also studied on the plasmodium of *Didymium nigripes* both with respect to bioelectric potential (WATANABE, KODATI and KINOSHITA 1939 a, b) and taxis (KINOSHITA 1939). It was found that the

---

[7] It was later shown by SHOWACER and DUBUY (1947) that infiltration of *Avena* coleoptile sections does not prevent the acceleration of elongation by IAA if an adequate oxygen supply is present.

plasmodium (*Dydimium nigripes*) shows a positive chemotaxis toward IAA in the concentration range between 0.04–0.63 mg./l. (KINOSHITA 1939); if the concentration is raised to 5 mg./l., positive chemotaxis is reversed to negative. Since a plasmodium owes its locomotion to protoplasmic streaming, we may suppose that IAA must have a profound influence on the protoplasmic streaming in a gymnoplast as well.

A characteristic feature of the auxin effects in both *Avena* and *Tradescantia* is that the acceleration is only transient, for the streaming rate

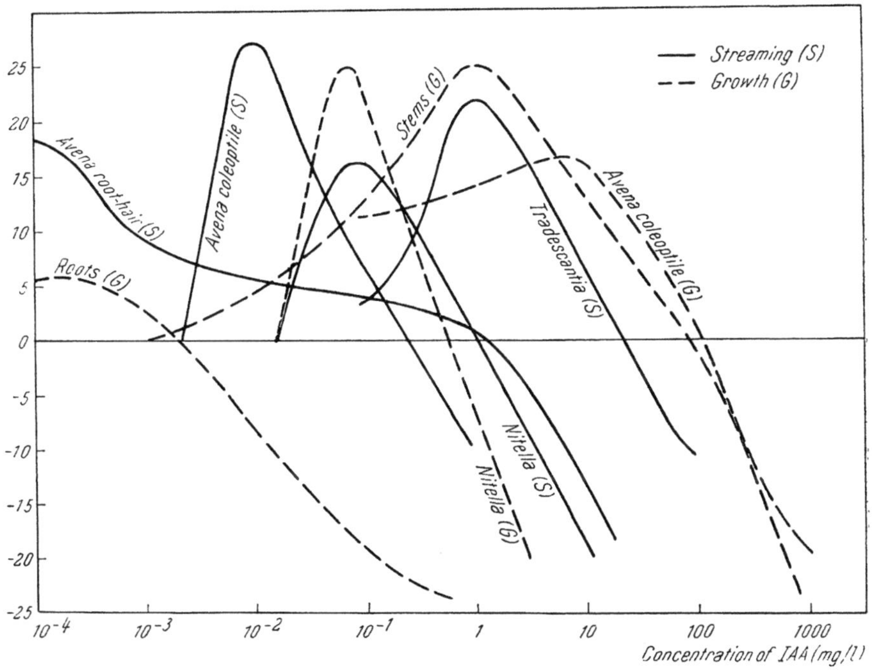

Fig. 63. Concentration ranges over which IAA exerts its effects on growth and on protoplasmic streaming in different plant organs. Ordinates: increase or decrease in growth or in rate of streaming; the scale for the ordinates is arbitrary, and comparisons of percentage effects are not attempted.
[After KELSO and TURNER (1955) with the addition of SANDAN and OGURA'S data (1957).]

returns to that of the control within 1 hour after the first application of auxin, although fresh auxin is being continually applied. The accelerating effect appears immediately and is over within 30 minutes for *Avena* (THIMANN and SWEENEY 1937 a, b) and 20–60 minutes for *Tradescantia* (KELSO and TURNER 1955). As seen from continual records of the rate of flow, the effect of a low concentration of IAA (0.01 mg./l.) reaches its maximum (20 μ/sec.) 10–20 minutes after the auxin is applied; then the streaming slows down till the initial value (17 μ/sec.) is reached after 30 minutes (cf. SWEENEY and THIMANN 1938, Figs. 1–3). It is noteworthy that recovery from stimulation takes the same time as recovery from inhibition brought about by a high concentration (higher than 0.5 mg./l. for *Avena*; 50 mg./l.–100 mg./l. for *Tradescantia* hair).

Sweeney and Thimann (1938) showed that a second stimulation by auxin can be obtained if the cells are washed with water and are left in it for a period of 30 minutes, or, alternatively, if fructose or some other sugar is added with auxin. They accounted for the transient nature of the auxin effect as due to the exhaustion of a substrate. This view is supported by a fact found by them that the initial auxin effect was no longer transient but was maintained if carbohydrate or malate (Sweeney and Thimann 1942) was added with the auxin. The work of Kelso and Turner (1955) on *Tradescantia* hair further confirms the above view. Fructose (0.03 M) or malic acid (0.01–0.1, pH 4) alone has no effect, but when added with IAA or indolebutyric acid (IBA), they modify the auxin effect on the streaming in the same way as in the case of *Avena* coleoptile, namely, they stabilize the effect for each concentration of auxin no matter whether it is stimulation or inhibition.

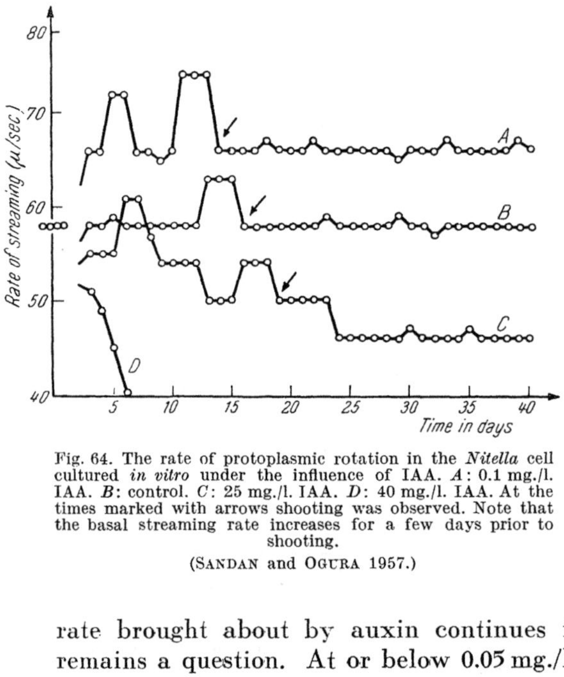

Fig. 64. The rate of protoplasmic rotation in the *Nitella* cell cultured *in vitro* under the influence of IAA. *A*: 0.1 mg./l. IAA. *B*: control. *C*: 25 mg./l. IAA. *D*: 40 mg./l. IAA. At the times marked with arrows shooting was observed. Note that the basal streaming rate increases for a few days prior to shooting.

(Sandan and Ogura 1957.)

In contrast to the above, the effect of auxin on the protoplasmic streaming in *Nitella* has a rather different aspect. At the optimum concentration (0.1–0.2 mg./l.), the rate of streaming, which is about 115% of the control, is maintained for 40 days or even more without the addition of any sugar or organic acid (Sandan and Ogura 1957). Just how the augmented rate brought about by auxin continues for such a long period of time remains a question. At or below 0.05 mg./l., the rate is not affected; in the concentration range 5.0–30.0 mg./l., the rate is not kept at a definite level. An experiment using 25 mg./l. showed that the rate eventually became stabilized after 24 days at a definite level (46 μ/sec.), which is about 80% of the control (58 μ/sec.) (Fig. 64). It may be that there are sufficient substrates in the cell which make it unnecessary to add substrates from without for maintaining the augmented speed.

The effects of IAA on the protoplasmic streaming is also shared by other auxins. Thimann and Sweeney (1937 b) showed that both coumaryl-3-acetic and allocinnamic acids, which have a growth-promoting activity from 1/10 to 1/100 of IAA, also cause an immediate increase in the rate of flow in the *Avena* coleoptile; this effect disappears within about 30 minutes. The optimal concentrations of the flow-accelerating effect of these two auxins are both 10 times that of IAA, which would be expected from their activity as auxins. Using intact *Oryza* seedlings, Doi (1950) observed that the streaming in the root hairs is stimulated with α-naphthaleneacetic acid (NAA) at a concentration of 0.1 mg./l., which nearly coincides with the optimal concentration for root formation. KELSO and TURNER (1955) compared the effect of indolebutyric acid (IBA), indoleacetonitrile, α-naphthaleneacetic acid (NAA), and 2,4-dichlorophenoxyacetic acid (2,4-D) on the streaming in *Tradescantia* hair with that produced by IAA. These growth-promoting substances bring about either acceleration or inhibition of the flow according to the concentration. The maximum effects appear in 30 minutes and the altered rates go back to the normal in 60—70 minutes except in the case of NAA (high concentrations) and 2,4-D (low concentrations).

It is still uncertain as to how the streaming is accelerated by auxins. As the change in the rate of flow must be considered in terms of both motive force and viscosity, it is significant to know if auxins may affect the protoplasmic viscosity appreciably at as low a concentration as that effective for stimulating the streaming. It is known that the viscosity of protoplasm in *Elodea* cells is not much altered by IAA up to a concentration of 1 mg./l. for at least 10 hours (STÅLFELT 1947/49). If we admit that this is also true for epidermis cells of the *Avena* coleoptile, the acceleration in *Avena* caused by IAA must be due to an increase of the motive force.

COMMONER and THIMANN (1941) and BERGER, SMITH and AVERY (1946) showed that auxin increases respiration of the *Avena* coleoptile if a suitable substance such as sugar, malate or fumarate are added with it. NAGAO and OHWAKI (1954) reported, however, that IAA (1 mg./l.) accelerates fermentation rather than respiration in the seedlings of *Oryza sativa* submerged in solutions of various oxygen concentrations. That the auxin promotes respiration or, according to circumstances, fermentation (NAGAO and OHWAKI 1954) may probably have to do with the fact that auxin action is involved in phosphate transfer (BONNER 1949, BONNER and BANDURSKI 1952), and phosphorylation in turn is supposed to have a direct bearing upon protoplasmic streaming as well as growth. However, nothing definite about the mechanism of the flow-promoting action of auxin is yet known. Just as the problem of the growth-promoting action of auxin is still undecided, thus also the mechanism of acceleration of the protoplasmic streaming is still an unsolved problem.

17. Fat solvents

Fat solvents, *e. g.*, alcohol, ether, chloroform, urethane, etc. are surface-active and some of them have a remarkable anaesthetic action on animals. It is known in the old literature that they also exert a pronounced effect on protoplasmic streaming. Plants often used for the experiments were *Chara, Nitella, Elodea, Vallisneria,* etc. It was observed by several authors that ether, chloroform etc. accelerate the protoplasmic streaming when the concentration is low but retard or stop it at higher concentrations (cf. KLEMM 1895, JOSING 1901, EWART 1903, etc.). On the other hand, BÄRLUND (1938) reported that the protoplasmic rotation in *Chara* cells was always slowed down with 1–2.5% ether reversibly, no acceleration being observed at any concentration. OLSEN and DUBUY's observation (1940) in *Avena* shows the same thing, namely, that the protoplasmic streaming is retarded with ether solution reversibly at concentrations of 1–3%; below 1% they could detect neither increase nor decrease in streaming rate. Also with chloroform and with alcohol no increase in the rate of flow could be noted.

SANDAN and YAMADA (unpublished) studied the effect of a series of monohydric alcohols as well as ether, chloroform and acetone upon protoplasmic rotation in *Chara* internode. After the external solution (distilled water) is replaced with a solution containing a fat solvent, the rate of streaming decreases once conspicuously, but subsequently turns into a stabilized rate characteristic of the kind and concentration of the fat solvent used. The period in which the rate is kept constant at a definite level lasts usually many hours. When water is substituted for the solution, the original rate is quickly restored.

Table 5. *Relation between concentrations of ethanol and the rate of flow in Chara Braunii (temp. 18.5° C., unpublished data of* SANDAN and YAMADA*)*.

Concentration (%)	Rate of flow (μ/sec.)	Rate in per cent of control
	81.4	100
1.0	71.2	87.3
2.0	63.3	77.7
3.0	54.2	66.6
4.0	47.5	58.4
5.0	40.7	50.0
6.0	33.5	41.1
8.0	0	

As the rate is stabilized at a definite level under solutions of most of the fat solvents of a definite concentration, we can find a rather exact quantitative relation between the concentration of a given fat solvent and the rate of flow in it. No acceleration of flow could be found by SANDAN and YAMADA at any concentration whatever of various alcohols, ethyl ether, chloroform and acetone.

According to SANDAN and YAMADA the relationship between the concentration of alcohols (methanol, ethanol, propanol, butanol) and the stabilized rate of streaming is nearly linear. Table 5 shows the case of ethanol for example. A similiar linear relationship holds true also for chloroform and ether, but the linearity between the concentration and rate of flow is not always found in other fat solvents, *e. g.*, acetone. It was also shown that the inhibiting effect of the streaming of monohydric alcohols increases with the number of carbon atoms, a fact which is pointed out by NOTHMANN-ZUCKERKANDL (1915) on the basis of the time necessary to stop the streaming in *Vallisneria*. She found that the toxicity of the homologous alcohols increases approximately corresponding to TRAUBE's law, but she also found that the surface activity of the reagents is not parallel to the flow-inhibiting effect.

Urethane also has a marked effect on the rate of streaming in *Nitella*. SIBAOKA and ODA (1956) found that the cell, whose streaming is retarded to a considerable extent under the influence of a 1% aqueous solution of ethyl urethane, can still be the source of an action current which is accompanied by instantaneous stoppage of streaming.

NICHOLS (1930) studied the effect of chloroform and ether when they are admitted only to a small spot on the *Nitella* internode by means of a small pipette. The streaming of protoplasm ceases at the point which is in direct contact with the tip of the pipette filled with chloroform and at the region downstream, but not in the other loci of the cell. When the granules in the moving protoplasm come against a quiet chloroformed spot, the result is

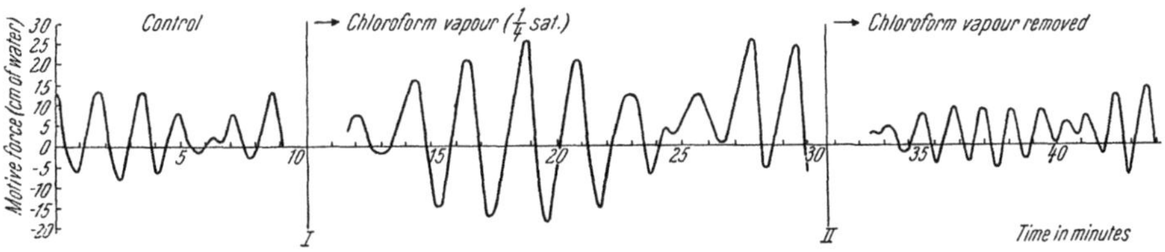

Fig. 65. Generation of the motive force *(Physarum polycephalum)* under the influence of diluted chloroform vapour (chloroform-saturated air diluted four times). (OHTA 1952 a.)

similar to that seen when the cell is punctured (NICHOLS 1925), namely, the movement of the granules is stopped, while those a little to one side do not stop completely and can go slowly around the sides of the affected area, eventually joining the normally moving stream beyond.

As for the myxomycete plasmodium, VOUK (1913) observed that ether upsets the rhythm completely. SEIFRIZ (1941) studied the behaviour of *Physarum* plasmodium under the effect of several anaesthetic agents. He found that chloroform and acetone are superior to ether with respect to the flow-inhibiting effect and its reversibility.

It was often supposed that the possible acceleration, retardation or cessation might primarily reflect the change in viscosity caused by the action of fat solvents, since there are some few data in favour of this view (cf. HEILBRUNN 1928, 1952). What is necessary for us in order to answer the question definitely is, however, to know how the motive force back of the streaming behaves under the influence of fat solvents. Fortunately, as far as the plasmodium is concerned, we are now in a position to investigate it with the aid of the double-chamber method. Among fat solvents, chloroform (OHTA 1952 a) and ethyl ether (KAMIYA 1953, KAMIYA, NAKAJIMA and ABE 1957) have been so far studied in connection with their effect on the production of the motive force.

OHTA pointed out that the effect of chloroform vapour on the production of the motive force is different according to the concentration and to some extent also to the individual specimen. Chloroform tends to enhance the motive force when its concentration is lower and to diminish it and upset the periodicity when its concentration is higher (Figs. 65 and 66). We notice that the DPG shown in Fig. 65 bears a close resemblance to that obtained when the respiration of the plasmodium is suppressed by low oxygen tension or respiratory inhibitors (cf. Figs. 39 and 41). As a matter of fact there is a hardly recognizable difference in DPG patterns between the two. It is not improbable that the increase in augmentation of the motive force in diluted chloroform vapour is as a matter of fact due to inhibition of respiration rather than to a direct effect on the driving mechanism. The fact that the motive force increases rather than decreases when respiration is suppressed, and the possible reason for this seemingly paradoxical relation have been already described and discussed (see pp. 66 and 74). The cause of decrease in the motive force under the effect of higher concentration is probably a matter of the mechanism generating the mechanical force.

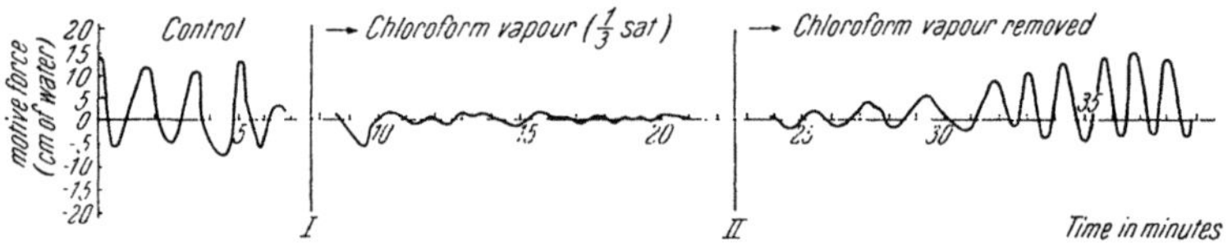

Fig. 66. Generation of the motive force *(Physarum polycephalum)* under the influence of chloroform vapour with a higher concentration (chloroform-saturated air diluted three times). (OHTA 1952 a.)

Even in this case the endoplasm is not gelated yet. At any rate, OHTA's results show that the action of fat solvents on the protoplasmic streaming can be very different according to the concentration, and that the effect must be considered not only in terms of viscosity but also of the motive force.

In the case of ether, too, the motive force is not much diminished if the concentration is low, although periodicity is strikingly affected (KAMIYA, NAKAJIMA and ABE 1957). When the concentration gets higher (1% aqueous solution), however, the motive force disappears almost completely (KAMIYA 1953).

It is to be stressed that all these reactions of the plasmodium, either to chloroform or to ether, are perfectly reversible without being accompanied with any observable ill effect.

18. Miscellaneous substances

Besides the influences of physical and chemical agents described in the foregoing pages, the effects of diverse substances were also the subject of study in connection with the protoplasmic streaming. There are but few substances which do not affect the rate of flow if they enter the cell at all. Therefore simple observation of the changes in the rate under the effect of arbitrarity chosen substances has no great meaning unless it leads to further understanding. We shall consider in this chapter the effect of some of the substances which were not considered in previous chapters.

(1) *SH-reagents.* In order to investigate the meaning of the sulfhydryl group for protoplasmic streaming, the effect of SH-reagents, which block the

sulfhydryl group, is of special interest. According to SMIRNOVA (1955, see also KOSHTOYANTZ 1957) rotational streaming in *Elodea* is retarded by the application of 0.1% ($= 5.5 \times 10^{-3}$ M) $CdCl_2$ and is eventually brought to a complete standstill. The flow, however, is always resumed at the initial rate when 1% ($= 8.3 \times 10^{-2}$ M) cysteine is admitted. Sometimes the protoplasm flows in the presence of cysteine with a rate even higher than before $CdCl_2$ is applied. This is probably due to stimulation by the —SH from the outside.

ABE (1959) studied the effect of para-chloromercuribenzoic acid (PCMB) on the *Nitella* cell. Upon admittance of 5×10^{-4} M PCMB the flow stops almost instantaneously. 10^{-2} M cysteine, however, restores the streaming. Such a recovery does not occur without cysteine. The stoppage of the flow by SH-reagents and the recovery of flow through cysteine point to the importance of the SH group in the protoplasmic streaming.

(2) *Ethylene chlorhydrin and thiourea.* Ethylene chlorhydrin and thiourea, both of which are known to be effective in breaking the rest period of freshly harvested potato tubers (DENNY 1926), are also potent in increasing the rate of protoplasmic streaming in *Nitella* and *Elodea* (MARCY 1937). After a 2 hour exposure, the maximum rates are brought about by ethylene chlorhydrin at a concentration of 1% in *Elodea*, and of 0.025% in *Nitella*; in the case of thiourea the maximum accelerating effect is obtained at a concentration of 0.05% in *Elodea* and 0.0075% in *Nitella*. The rate of flow of treated protoplasm as compared with that of untreated protoplasm is about twice the normal both in *Elodea* and *Nitella*. NORTHEN (1946) found, on the other hand, that the structural viscosity of *Elodea* protoplasm decreases considerably after 2 hours' immersion in both ethylene chlorhydrin (0.5–2%) and thiourea (1%). Thus so far as comparison is possible, it appears that the increase in the streaming rate and the decrease in the structural viscosity go in parallel. For comparing the data of the two authors, however, it is somewhat unfortunate that NORTHEN did not investigate the effect of much lower concentrations of the substances for longer periods of time, under which conditions the augmentation of the streaming rate is still very conspicuous. The promoting effect of ethylene chlorhydrin on the streaming is not seen in coleoptile cells of *Avena* (THIMANN and SWEENEY 1937).

(3) *Herbicides.* CURRIER (1949) studied the effect of four herbicides, 2,4-dichlorophenoxyacetic acid (2,4 D) 4,6-dinitro-2-sec. butyl phenol (DNSBP), pentachlorophenol (PCP) and O-isopropyl-N-phenyl carbamate (IPC) on outer epidermis of *Allium*, staminal hairs of *Tradescantia*, leaves of *Anacharis* (*Elodea*), etc. The effective concentrations were different for different drugs and for different cells. Generally speaking, however, the initial response is an increase in rate through changes in viscosity followed by a stage in which the rate decreases and the flow becomes jerky and erratic, and eventually by a stage of irreversible injury in which the streaming comes to a standstill.

(4) *Insecticides.* ITO and YAMAMOTO (1957) investigated in their cell-physiological studies on the influences of some insecticides the effect of

Parathion, Diazinon, Endrin and BHC on the rate of streaming in root hairs of *Oryza sativa* and *Triticum vulgare*. The minimum concentrations of the above substances needed to stop the streaming are smaller in *Triticum* than in *Oryza*, a fact which indicates the higher sensitivity of the former to the insecticide. In *Oryza* root hairs, the streaming was accelerated by Parathion at a dilution of 50.000× (9.4 p.p.m.) as compared with the control in distilled water, whereas no such effect was seen in wheat root hairs.

(5) *Alkaloids and snake venoms.* There are many old reports on the effect of alkaloids on the protoplasmic streaming in plant cells. They are surveyed by EWART (1903).

LEPOW (1938) studied the effect of alkaloids (cocaine, quinine, atropine, morphine, ephedrine, strychnine) and snake venoms (*Bothrops atrox, Crotalus atrox*) and some other substances in varying concentrations on various physical and colloidal states and also on protoplasmic streaming of *Physarum* plasmodium. He applied the substances to the plasmodium locally by means of a micropipette. The initial effect on streaming was a transient acceleration of the rate. The subsequent reaction was different according to the substances used. Quinine hydrochloride causes complete cessation of flow a few seconds after the initial stimulation. Morphine sulphate eliminated all motion in a few minutes. ENGEL (1954), who injected RUSSEL's viper venom (10^{-3}–10^{-4} dilution) containing lecithinase A into *Physarum* plasmodium, found that the gel matrix is decreased by the injection and the protoplasmic streaming is accelerated. As sodium stearate and sodium citrate induce similar changes in the activity of the plasmodium and $CaCl_2$ has the opposite effect, he interpreted the changes brought about by the snake venom, stearate and citrate as due to the removal of calcium from combination with proteins.

(6) *Dyes.* There are many older observations to the effect that protoplasmic streaming is maintained when the vacuole or even the cytoplasm is stained with various kinds of dyes. BEIKIRCH (1925) reported that, among many dyes which he used, methylene blue, neutral red and chrysoidine stimulate the streaming in *Elodea* cells. But OLSON and DUBUY (1940) found no acceleration of the streaming in *Avena* coleoptile with methylene blue (0.01 and 1 mg./l.).

As neutral red often forms innumerable minute precipitations in the *Elodea* vacuole, we can, by taking them as indices, clearly follow the streaming rate of the cell sap at various loci in the vacuole. GOLDACRE (1952) stained plant root hairs with neutral red and found that the dye accumulates at the end of the vacuole toward which the cytoplasm streams. It is instructive indeed to compare this phenomenon with his inversion tube analogy, but it is difficult to correlate this phenomenon directly with the cause of the streaming. Like the photograph presented by GOLDACRE (Fig. 8*b* in his paper), the vacuole is often disrupted by the formation of protoplasmic septa or bridges as a result of vacuole contraction. The phenomenon was observed in root hairs and other long cells in a variety of conditions, especially in acid solutions (STRUGGER 1926, 1928; SAKAMURA and

Kanemori 1935; Doi 1950). It is most impressively shown when the cell is strongly stained with neutral red. At any rate, when the septa are formed in a root hair, protoplasmic rotation usually slackens or ceases. Judging from the photograph, Goldacre's observation on the root hairs seems to be concerned with the state of vacuole contraction. Unfortunately, there are no descriptions about the material and conditions of his experiment on root hairs. A remarkable case in which cells in a state of strong vacuole contraction display a lively protoplasmic motion was observed by Weber (1930) in *Elodea* stained with neutral red.

Höfler, Ziegler and Luhan (1956) observed that the streaming protoplasm in root hairs of *Trianea* can be stained, although not strongly, with fluorescent dyes, such as uranin or rhodamin B.

(7) *Enzymes.* Gimesi and Pozár (1955) reported that the proteolytic enzymes cause an increase in the rate of protoplasmic streaming in *Elodea.* For instance, with 0.02% papain the rate increases 1.5 times the control and with 0.02% pepsin and digestive sap of *Nepenthes* the rate is augmented 2.1 and 1.8 times the control, respectively. It is probable that free amino acids produced by proteolytic enzymes are responsible for the stimulation of the streaming.

Haas (1953) injected into plasmodia of *Physarum polycephalum* seven enzymes at various concentrations. They were: papain, trypsin, pepsin, collagenase, pectinase, hyaluronidase and ribonuclease. The observation was qualitative rather than quantitative and mainly concerned visible changes in morphological, colloidal and motile aspects of the plasmodium. The common property of the proteinases is that they accelerate protoplasmic streaming and produce numerous hyaline protuberances. The polysaccharases, pectinase and hyaluronidase produce an increase in the rate of protoplasmic streaming but depress the forward motion of the organism as a whole. Ribonuclease acts in a similar way. Pectinase, hyaluronidase and ribonuclease promote the solation of plasmagel.

Although the interpretation of the results involves difficulties because of their variability and complexity, Haas drew tentative conclusions that the hyaline layer of the plasmodium consists mainly of non-conjugated proteins whereas the plasmagel contains proteins in combination with other substances such as polysaccharides and nucleic acid.

VIII. Biology and Function of Protoplasmic Streaming

1. Inducement of protoplasmic streaming

It has long since been known that in many plant cells such as *Elodea* or *Vallisneria,* protoplasmic streaming is initiated by external influences (Frank 1872, Keller 1890, Hauptfleisch 1892, Kretzschmar 1904, etc.). This "induced" or "secondary" streaming is distinguished from "spontaneous" or "primary" streaming as it occurs in *Chara* or *Nitella* cells, many root hair cells etc.

When a leaf is torn off from the plant, a lively protoplasmic streaming is initiated in it after a certain period of latency, a movement that was not observable when the plant was not injured. We do not yet know the meaning nor the purpose of this movement. It is therefore not without reason that it is considered as a "merely pathological" phenomenon. But, rather than discuss whether it is pathological or not, we feel more interested in the fact itself and in the question as to what the change in situation is that forms the factor initiating this movement.

Much work has been done with the purpose of analysing the nature of stimulus producing the protoplasmic flow. It was found that besides injuries, changes in illumination or in temperature, chemical agents etc. are also effective in inducing the streaming. In all the experiments dealing with the inducement of the protoplasmic flow, it was pointed out that there is always a certain lapse of time after the stimulus has been applied, before the streaming is brought about. Soon after the preparation of, say, an *Elodea* leaf, the rotational streaming is not observed; it begins at the base of the leaf adjacent to the region of injury and spreads further from it gradually. The motion of protoplasm in this short lapse of time, or in the pre-stage of the streaming, which lasts usually around 10 minutes, is described as irregular motions of microsomes by DEHNEKE (1881), WIGAND (1885), HAUPTFLEISCH (1892), FITTING (1925), etc.

a) Chemodinesis

FITTING (1925—1937) has published many detailed studies on external conditions inducing protoplasmic streaming. Most of his experiments were conducted on *Vallisneria spiralis*. He termed the inducement of the streaming by means of chemical agents "chemodinesis."

It was FITTING (1925) who first noted that the streaming is brought about by hormones produced when injuries are given. He found that the extract of *Vallisneria* leaves is effective in causing protoplasmic streaming even in dilutions of 0.03–0.008%, *i. e.*, an extract prepared from 1 gm. fresh leaf with 3–13 l. water. The substance in question is very stable to heat but is very easily destroyed by bacteria and is not volatile. Therefore it is readily obtained by heating the leaves in water. It is also effective in other kinds of plants; for instance the extract of leaves of *Vallisneria* has an effect on *Elodea* and that of *Elodea* upon *Vallisneria.* Moreover it was found that the water extract of filter paper has the power to initiate protoplasmic streaming in *Vallisneria* and *Elodea* and that the same can be said of copper ion as well.

Moreover by experimenting with various substances, it was found that α-amino acids, which are the products of the decomposition of protein, have a very strong action as compared with other substances. Among amino-acids, especially L-histidine and L-methylhistidine show an action at a concentration as low as 10^{-8} M. A somewhat higher concentration is necessary for L-glutamic acid, L-aspartic acid, L-phenylalanine, L-serine, L-alanine, etc. Generally speaking, the minimum effective concentrations of these amino acids are extremely low as compared with those of other substances, but some of the amino-acids, *e. g.*, arginine, lysine, cystine, tyrosine, tryptophan, etc., are known to have no effect at all (FITTING 1927).

Especially to be noted is the fact that optic isomers are not identical in their effect (FITTING 1929). For instance, the threshold for D(+)-histidine is 30—50 times higher than for L(—)-histidine and D(+)-aspartic acid is even (50—) 200 times less effective than L(—)-aspartic acid. A still greater difference in sensitivity is found for D(—)- and L(+)-alanine; the former needs (400—) 1000 times as high a concentration as the latter. This difference in the effect of D- and L-components is intelligible only on the assumption of an asymmetric structure of the protoplasm.

The reaction of the protoplasm to the α-amino acids is only transient. A solution of an amino acid at a definite concentration brings forth protoplasmic streaming only for a definite time, after which the streaming ceases. Therefore in order to initiate the protoplasmic streaming again by applying the same substance to the same material, it is necessary to administer a solution with a concentration from 2.5 to 4 times as high as that of the former solution (Unterschiedsschwelle). That is, the effect of this substance diminishes with time (Abstumpfung). The material to which a substance (an amino acid) in a definite concentration has become no longer effective can be activated by means of a second or third substance (different kinds of amino acids). In this case, the concentration needed for activation is different from that required for the one and the same substance to be effective at the second time. That is to say, it is considered that each substance has a different action upon the protoplasmic streaming. Thus we may assume that the protoplasm can by itself distinguish different substances. By using this characteristic of the protoplasm, it has been observed that the extract of the leaf and L-histidine act upon the protoplasmic streaming as if they were entirely the same kind of substance. Other experiments show that there is in the living leaf of *Vallisneria* a nonvolatile substance which is stable to heat and which, diffusing out to the surrounding water, lowers the sensitivity to amino acids. Consequently this seems to be the effective substance found in the extract of the leaf.

Besides amino acids various substances are known to be capable of inducing the protoplasmic streaming. That IAA is also effective as an inducing agent, was also found by FITTING (1937), but it is not as effective as amino acids since the threshold concentration initiating the streaming is somewhat higher than 10^{-4} M, *i. e.*, 18 mg./l. This concentration is much too high as compared with that bringing about the maximum acceleration of streaming rate in *Avena* coleoptile. It is also higher than the optimum concentration for the streaming rate in *Tradescantia* hair or *Nitella* internode (cf. Fig. 63, p. 121).

It is known that protoplasmic streaming is also induced by the salts of some heavy metals, such as manganese, magnesium (BEIKIRCH 1925) and copper (SEIFRIZ 1922, FITTING 1925). The inducing effect in this case is very conspicuous, but the phenomenon is obviously pathological; the streaming continues furiously for days. Besides sulfuric acid (0.005%) (LAKON 1914), methanol (3–10%) and saponin (1–2%) (SEIFRIZ 1922), etc. are also effective.

We thus know that from substances like histidine, which is effective at a concentration as low as 10^{-8} M, to those like methanol or saponin, which require a concentration as high as a few percent, various kinds of agents have the ability to induce the streaming. The underlying mechanism of the initiation of the streaming is as yet far from clear, but it may well be supposed that it has something to do with the shift in the metabolic pathway and in the energy-yielding reaction directly connected with protoplasmic streaming.

Protoplasmic streaming in the coleoptile of *Avena*, especially when it is young, is known to suffer retardation due to oxygen deficiency (p. 65). SWEENEY and THIMANN 1938 found that this retardation is delayed by very dilute solutions (6×10^{-7} M) of histidine. They, therefore, believed that histidine must reduce the respiration of the *Avena* coleoptile, and further assumed that the metabolism, whose energy is readily available for the streaming, is augmented by the suppression of the respiration which is not available for the streaming.

The effect of histidine may also be connected with its metal-chelating action FUJII *et al.* (1955) found the interesting fact that the motility of starfish spermatozoa, which are generally immobile, was instantly activated upon addition of histidine, cysteine and glycine to their suspension. Some chelating agents, *i.e.*, ethylenediaminetetraacetic acid (EDTA) or diethyldithiocarbamate (DDC) also have the same effect. They suggested that the initiation of this activity is due to the release of zinc from the surface of the spermatozoa.

Recently OHTA (1958) investigated the effect of various amino acids of α-L-type on the generation of the motive force for the protoplasmic streaming in the slime mould, *Physarum polycephalum*, by means of the double-chamber method. The amino acids tested by him were histidine, cysteine, glycine, alanine, leucine and glutamic acid. Of these amino acids, histidine exerts again a remarkable influence. When the whole plasmodium is treated with histidine at a concentration from 10^{-2} to 10^{-3} M, the motive force increases conspicuously. Cysteine augments the motive force in the same way as histidine, but the effective concentration (10^{-2}–2×10^{-3} M) is somewhat higher than that of histidine. Glycine is less effective (effective concentration: 5×10^{-2}–10^{-2} M), and alanine, leucine and glutamic acid have no effect.

OHTA assumed that the stimulation of the motive force may, in analogy to the case of the starfish spermatozoa, also be due to a release of metal (probably zinc) from the plasmodium. An evidence for this is the fact that histidine solution, in which the plasmodium had been suspended for 30 minutes, always gave a positive dithizone reaction. Positive dithizone reaction was also observed in the case of cysteine and glycine both of which stimulated the motive force, but it was negative for the other amino acids which had no effect on the motive force. Whether the chemodinetic effect studied by FITTING has also to do with the release of metal by amino acids is yet unknown, but OHTA's results are interesting when compared with FITTING's work.

b) Photodinesis

Inducement of the streaming through light has also been known for a long time. It was MOORE (1888) who first observed the initiation of the protoplasmic streaming through light in *Vallisneria* and *Elodea* leaves. Later the stimulating effect of light on protoplasmic streaming has been referred to as "photodinesis" (FITTING 1925, p. 371).

NOTHMANN-ZUCKERKANDL (1915) studied in *Elodea* the photodinetic effect of various wave lengths. She found that intense light from the ultraviolet to the infrared is capable of arousing the streaming and that the light of longer wave length is more effective. That this is simply due to heat action is not likely because applied heat, according to NOTHMANN-ZUCKERKANDL, does not always induce active streaming.

BEIKIRCH (1925), who also studied the light effect using *Elodea* cells, came to a conclusion similar to that of NOTHMANN-ZUCKERKANDL. He claimed that the stimulating effect of light is brought forth not by accelerating the rate of flow but rather by initiating streaming in quiescent cells. Thus the number of cells showing protoplasmic streaming is increased by light. According to FITTING (1925), protoplasmic rotation in the *Vallisneria* leaf is induced through transition from a weak light to a stronger one. The inducing factor is not the light intensity but the change from a weaker to a stronger light. The material which has been adapted to darkness is particularly sensitive to light and the rotation is initiated by illumination lasting only a few minutes.

SCHWEICKERDT (1928) studied the photodinetic effect in detail in the *Vallisneria* leaf. Light-sensitivity of the dark-adapted *Vallisneria* leaf is very great. For 22.2 lux, 80 seconds are sufficient; for 2560 lux only 4.5 seconds are enough. Even under as low a light intensity as 0.5 lux, streaming is evoked in a dark-adapted cell with a sufficiently long period of illumination. Exposure time necessary to provoke the streaming is dependent on the light intensity; the weaker the light the longer the duration of illumination. Also the reaction time depends on the light intensity, although at least 5 minutes are necessary at higher intensities. The reaction time is therefore much longer than the duration of exposure to light. Also for the same light quantity, the stimulating effect is greater under lower intensity than under higher intensity. The product law ($i \,.\, t = c$, where i: intensity, t: time, c: a constant) does not hold for SCHWEICKERDT's results. Intermittent, subthreshold stimuli act in summation if the ratio of light period and dark period is 1 : 1 or 1 : 2. Continuous exposure to light of high intensity (800 lux and over) brings about damage to the cell.

In contrast to the result of NOTHMANN-ZUCKERKANDL and BEIKIRCH, SCHWEICKERDT's experiments show that infrared radiation is not capable of inducing protoplasmic streaming. Red, green and blue lights are effective for photodinesis. Red light is most effective, then blue, green being least effective. Continuous illumination of red or green light does not maintain the streaming which is evoked with the same light. Rotation in red or green light stops completely in the course of days. In white light (even of only 0.5 lux) or blue light, on the other hand, the protoplasmic streaming, which has been induced, does not stop completely. Thus red and green light act only as transient stimuli, whereas blue and white light act both as stronger transient stimuli and as weak stationary ones.

As already mentioned, NOTHMANN-ZUCKERKANDL (1915) reported that a proper dose of ultraviolet light is effective in inducing protoplasmic streamin in the *Elodea* leaf. Later the photodinetic action of ultraviolet rays was studied more fully by NOETHLING and ROCHLIN (1931). According to them ultraviolet radiation of a wave length around 300 mμ or shorter is especially effective not only in inducing the streaming but also in causing necrosis, which is characterized by cessation of movement and formation of oxalate crystals.

That the short wave ultraviolet rays affect the streaming in a manner different from visible light is known from the fact that, in the former case, cells which are kept in darkness near the radiated cells, or even cells in another leaf in the proximity of radiated cells, are also affected (cf. BÜNNING 1953). This indirect influence results in an acceleration of the streaming. It is probable that the cells, which are directly irradiated, produce substances capable of promoting the streaming in neighbouring cells and that these substances may be amino acids produced as a result of decomposition of proteins.

Perhaps the intracellular migration of chloroplasts in response to the direction and intensity of light stands in close relation to the phenomenon of photodinesis. It is well known that chloroplasts are arranged in a plane perpendicular to the direction of incident light (epistroph). This fact is known especially well in mosses, fern prothallia, *Elodea, Lemna,* etc. Under very intense light chloroplasts on side walls take a position parallel to the incident light (parastroph); also in the dark they wander to the side and inner walls (apostroph). ZURZYCKA (1951) investigated the relation of the wave length and the intensity of light to the movement of chloroplasts in *Lemna triscula* and found that the reaction from apostroph to epistroph is caused entirely by light absorption of carotenes and that from epistroph to parastroph by light absorption of chlorophyll. How the movement of chloroplasts occurring in response to light is causally related with the protoplasmic streaming is a classical and yet unsettled problem (cf. FRANK 1872, SENN 1908). A series of recent contributions of ZURZYCKI and ZURZYCKA (*e. g.,* ZURZYCKA and ZURZYKI 1957) however, have thrown important light on this problem.

2. Transport of substances

It has been often claimed that protoplasmic streaming may favour the transport of food materials from one part of a plant to another. The data so far obtained in different plants with different methods, however, are not always in conformity. Therefore the opinions among various authors as to the part played by protoplasmic streaming for the translocation of substances are still controversial.

That the protoplasmic streaming aids the transport of the substances was claimed by DE VRIES as early as 1885. This idea was supported later by many workers (KIENITS-GERLOFF 1893, JANSE 1890, BIERBERG 1909, etc.). DE VRIES (1885) and CURTIS (1935) insisted that transportation of nutrition through the sieve tube is also aided by the protoplasmic streaming. Recently, LOU, WU, CHANG and SHAO (1957) showed the existence of intercellular movement of protoplasm *via* the plasmodesmata in the parenchyma of various organs of *Allium sativum*. They endowed this cell-to-cell movement of protoplasm with the function of translocating not only low-molecular substances, such as sugars, amino acids, inorganic salts, etc., but also the partially disorganized protoplasm which still retains the complicated structures of a biocolloid.

When some three decades ago the importance of the role of growth hormone excited keen attention among plant physiologists, the mechanism of its transport also became a problem of great concern. In order to account for the rapid translocation of auxins through the *Avena* coleoptile, WENT (1928) assumed that protoplasmic streaming is responsible for it. The facts supporting WENT's view were presented by DUBUY and OLSON (1938) who showed a close interrelation between protoplasmic streaming, auxin transport and elongation in the coleoptile of oats and wheat. The rate of each of these activities is inhibited by the passage of a weak electric current through non-polarizable electrodes. These authors showed further (DUBUY and OLSON 1940) that all three processes, protoplasmic streaming, respiration and transport are limited to the same extent by the same concentration of KCN. Thus they supposed that the transport is controlled by protoplasmic streaming, and the streaming is regulated in turn by respiration, and hence any factor affecting respiration should likewise affect transport *via* streaming.

On the other hand, there are data which do not fit into the foregoing view. KOK (1931, 1933) studied the influence of protoplasmic rotation on the transport of lithium nitrate and caffein through *Vallisneria* leaves, but she was unable to ascertain any accelerating effect of rotation on transport. According to CLARK (1938), sodium glycocholate in a nontoxic concentration (1 : 500) completely abolishes auxin transport in *Avena* coleoptile without stopping the protoplasmic streaming. Saponin (0.05 saturated), on the contrary, stops protoplasmic streaming, but has no effect on transport. Although OLSON and DUBUY (1940) insist that saponin has no effect on the streaming rate at concentrations between 100 mg./l. and 100,000 mg./l., the experiments of CLARK point to the fact that auxin transport and protoplasmic streaming are not so closely related. The fact known in *Avena* coleoptile that electrotropic bending [8] is dependent on the direction of the applied current while protoplasmic streaming is apparently not (DUBUY and OLSON 1938) is also not consistent with the view correlating protoplasmic streaming with transport. A further fact showing a poor correlation between the streaming

[8] The bending caused by the current applied either longitudinally (SCHRANK 1949) or transversely (SCHRANK 1947, 1948).

and transport is to be had in the work of SCHUMACHER (1936) who found that the transport of fluorescein is independent of protoplasmic streaming in the hair of *Cucurbita Pepo.*

All in all, the role of rotation or circulation played in the translocation of soluble substances is still uncertain.

The situation, however, is quite different in coenocytic tubes of *Siphonales* or in hyphae of *Phycomycetes* where protoplasm itself is translocated. In *Caulerpa prolifera,* for instance, unfavourable conditions for the plant, such as streaming water, sublethal dosage of poisons, transient exposure to the air and other injurious effects induce the mobilization of protoplasm, that is, a part of the protoplasmic contents wanders out of the leaves and slowly streams to the rhizomes or to the lowermost leaf bases. DOSTAL (1929), who studied these phenomena in detail, distinguished two kinds of protoplasm in the *Caulerpa* cell: one is "stable," devoid of chloroplasts, and always remains at its own place, and the other is "unstable," carrying chloroplasts and ready to be mobilized in case of emergency. In normal leaves the latter adheres to the wall side of the 'leaf,' or rests on supporting trabeculae characteristic of the material, or constitutes a thickening of the protoplasmic strands. In the case of mobilization, however, they move basipetally, either to rhizomes or petiole bases. On the other hand, when fruit formation starts, the "unstable" part of protoplasm carrying chloroplasts streams acropetally, and is transformed into zoospores. Obviously the protoplasmic streaming is associated with reproduction in this case.

SCHRÖTER, who studied *Mucor stolonifer* and *Phycomyces,* correlates the growth of hyphae with the streaming direction. The strongest growth occurs when the protoplasm moves toward the tip. Protoplasmic streaming must be the means of transportation of the materials required for hyphae growth in this case, since if protoplasm is quiescent, the growth becomes insignificant; growth is hardly perceptible as soon as the stream hastens away from the tip. GREHN (1932) interpreted the mass streaming of protoplasm in hyphae of *Mucorineae,* in which streaming toward reproductive organs, fruit hyphae or zygotes is especially conspicuous. Incidentally, this mass movement is explained in terms of pressure streaming (MÜNCH 1930) caused by the difference in local osmotic pressure of the hyphal system or by transpiration.

In angiosperms, it is generally accepted that the transportation of the generative and vegetative nuclei along the pollen tube is occasioned by the protoplasmic streaming. It is clear that the protoplasmic streaming has an important bearing on fertilization in this case.

There is no doubt left as to the biological meaning of the protoplasmic streaming as a means of transportation in the above cases. Protoplasmic streaming plays here an obvious role with definite purposes.

3. Locomotion

In the slime mould and some amoebae the protoplasmic streaming and locomotion of the organism are so closely connected that one is hardly able to consider them separately. The locomotion of the plasmodium has been shown quantitatively by KAMIYA (1950 c) in terms of transport-volume (p. 35). It is obvious in these cases that no locomotion of the organism can take place without protoplasmic streaming.

As a special type of protoplasmic streaming we also notice one connected with the forward movements of *Pennatae* diatoms. MÜLLER (1896 a, b) showed the presence of peripheral strands of protoplasm which stream along the raphe and return by a central path in the body of the diatom. It is believed that this protoplasmic streaming, by functioning somewhat like a caterpillar mechanism, is responsible for the locomotion of the diatom. The friction of these bands against the substrate causes the diatom cell as a whole to move in the opposite direction. Recent contributions to the physiology of the movement of diatoms were made by the schools of KÜSTER (1956) and HÖFLER (1940 a).

4. Correlation with some physiological activities and seasonal variations

Augmentation of protoplasmic streaming as an indication of special performances of the cell or of preparation for them has been observed in various materials.

GEITLER (1942) described activation of the protoplasmic streaming in a cell of *Mesotaenium* after invasion by *Zygorhizidium verrucosum,* a member of *Phycomycetes.* TOMIYAMA (1954) observed, in his study on the processes of alteration produced by infection of *Phytophthore infestans* in the living epidermis cell of the midrib of young potato leaves, that the nucleus migrates toward the penetrating infecting hyphae. In the early stage of penetration, the cytoplasm is often seemingly thickened at the affected part, around which active protoplasmic strands appear and vigorous protoplasmic streaming becomes observable.

SANDAN (1955) observed in his experiment on the morphogenesis of an isolated internodal cell of *Nitella flexilis* that the rate of flow is increased by about 10% for several days prior to the shooting (cf. Fig. 64). The rate comes back to the original, constant level again when shooting begins. This kind of stimulation in protoplasmic flow may be an indication of the "increased activity" of the cell function at the preparatory period of shooting, but its causal relationship is so far unknown.

An example showing the correlation between the state of protoplasmic streaming and the change in the physiological condition of the cell under natural conditions is to be had in guard cells of a stoma. WEBER (1927 b) found in *Vicia* leaf that when a stoma is closed, a lively protoplasmic streaming is seen in the guard cells; when the stoma is open, however, the protoplasm shows either no motion or only local agitations. In the former case it is noticed that the chloroplasts remain at rest in spite of the active cytoplasmic streaming. WEBER's observation was confirmed by SINKE (1939) in various other plants.

It is often claimed that the basal rate of protoplasmic streaming in the same plant is sometimes different under the same experimental conditions according to the season of the year. SWEENEY and THIMANN (1938) noticed in their experiments using *Avena* coleoptile that the streaming rate of controls in pure water was 12.4 μ/sec. in November, whereas it was 17.6 μ/sec. in August. There was, however, no difference in rate according to the time of a day, a fact which was also ascertained by BOTTELIER (1934). From May to August, when the basal rate of flow is high, the effect of added auxin is less both at low and at high concentrations.

According to KELSO and TURNER (1955), the rate of streaming in the cell of the staminal hair of *Tradescantia* reached its minimum in midwinter (July at Melbourne, Australia) and showed two pronounced maxima, one in autumn (April-May), and one in spring (October-November). The maximum rate was about 7 μ/sec. and the minimum 6 μ/sec.

Also in the *Nitella* cell a distinct difference in behaviour in summer and winter was noted by BRECKHEIMER-BEYRICH (1954) as was stated before (p. 106).

IX. Experimental Approaches to the Mechanism of the Protoplasmic Streaming

There are apparently many means of approaching the study of the mechanism of the streaming. They are those offered by morphology, physiology, biochemistry, biophysics, etc. What is needed here for discovering the nature of the mechanism, is a synthesis of the many-sided data obtained by these different approaches. Here will be presented various facts that have been known up to the present and that seem to be directly important in considering the mechanism of the protoplasmic streaming—facts which of course are fragmentary—and we shall consider what can be inferred from them.

One of the basic questions concerning the mechanism of flow is to know in what part of the protoplasmic system the motive force responsible for it is produced. There has been a great divergence of opinion on this problem. The nature of the force responsible for the flow is another important problem. To begin with, we shall deal with the problem of intracellular location of the motive force and the mode of its application.

1. Seat of the motive force in rotational streaming

a) Deduction from the velocity distribution

It was shown in Chapter III that in a rhizoid cell as well as in a 'leaf' cell of *Nitella* the plasmasol (endoplasm) at different layers moves with a nearly equal rate (Figs. 5 and 6). In other words, there is only a negligible velocity gradient within the endoplasmic layer which is just sliding as a whole on the inner surface of the cortical gel layer. This type of velocity distribution seems to show that the active part in driving the endoplasm is the interface region between endoplasmic sol and cortical gel where sliding occurs and that endoplasm alone is rather inert in respect to the movement. We shall further consider this point in discussing the following experiment.

An appropriate operation combining centrifugation and subsequent ligation enables us to make a fragment of an internodal cell of *Nitella* or *Chara*. the whole space of which is filled with protoplasm without visible vacuole (Fig. 67). KAMIYA and KURODA (1956) observed a rather vigorous streaming of endoplasm in such a cell which had been deprived of its vacuoles by artificial means. In the light of this fact we can first of all know that the existence of the vacuole has no positive significance in the protoplasmic streaming. This is a decisive fact invalidating the view that the seat of the force causing the flow is located between the cell sap and the cytoplasm (BERTHOLD 1886, pp. 116, 122, cf. also LUNDEGÅRDH 1922, p. 379). Further, it has also been made clear that this intracellular streaming has a velocity distribution as shown in Fig. 68.

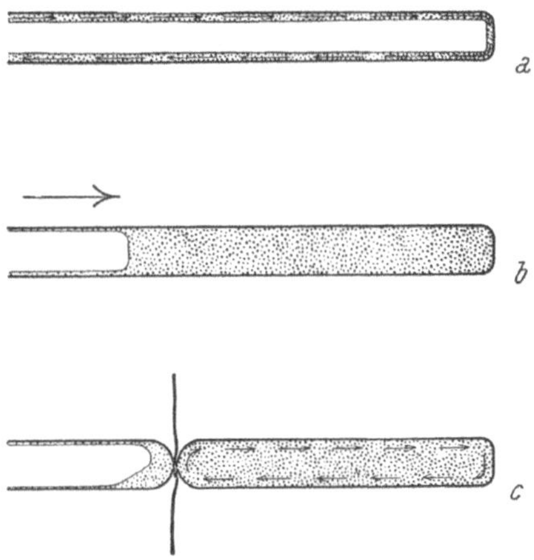

Fig. 67. The procedure for obtaining a plasm-filled cell fragment. *a* One end of an internodal cell of *Nitella* under natural condition. Chloroplasts in the cortical layer are not shown. *b* One end of the cell at which the endoplasm is collected under the influence of centrifugal force. Arrow indicates the direction of centrifugation. *a* A plasm-filled cell fragment produced through strangulation from the centrifuged cell.
(KAMIYA and KURODA 1956 b.)

In the cell of a normal rhizoid or leaf having a large vacuole, the sol layer of protoplasm slides over the surface of the gel as a whole almost without any velocity gradient occurring in itself as mentioned before, but the situation differs in a cell consisting solely of protoplasm. The velocity gradient here becomes greatest at the point of inflexion of the axial part (0.6 sec.$^{-1}$) and gets smaller as the peripheral part (0.1 sec.$^{-1}$) is approached. The speed, on the contrary, is zero at the axial part and becomes greater the nearer we get to the gel layer, but the greatest speed near the periphery (ca. 50 μ/sec.) suddenly drops to zero as the cortical gel is reached. The velocity curve is therefore almost discontinuous at both ends at the peripheries.

In order to understand the meaning of this velocity distribution, let us think of a long and round tube that has been made as a model imitating the dynamic conditions of a plasm-filled *Nitella* cell. That tube is supposed to consist of two semi-cylindrical walls that are placed one in front of the other (Fig. 69). When an appropriate fluid is placed in it and the two semi-cylindrical walls are slided against each other, how will the fluid in the tube be moved? Using warm margarine as a test fluid, KAMIYA and KURODA investigated this movement (velocity distribution) in reference to the section running through the axis of the tube and perpendicular to the surface of the shifting (corresponding to the optical section of the cell shown in Fig. 68). As is seen in Fig. 70, it was revealed that it bears a rather exact resemblance in form to the velocity distribution that is represented in a thick line in Fig. 68.

The fact that the protoplasm in a vacuole-free cell streams with an

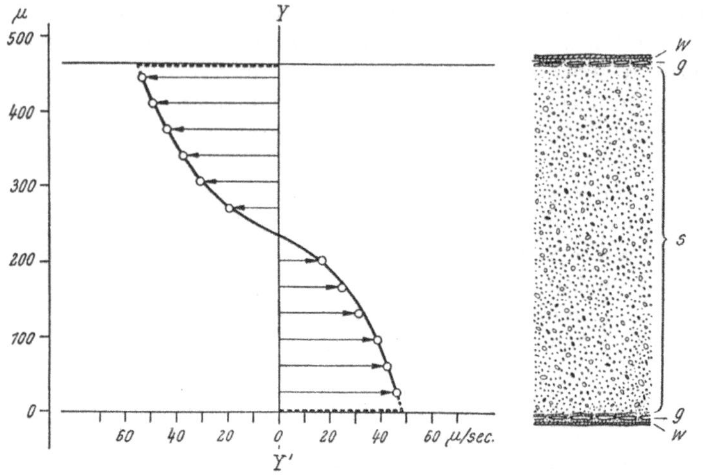

Fig. 68. Velocity distribution of the endoplasmic streaming in a cell, the entire space of which is filled with endoplasm. A part of the cell as seen under the microscope is shown on the right. *w*: cell wall, *g*: plasmagel layer in which chloroplasts are imbedded. *s*: plasmasol (endoplasm). No vacuole is visible. The rate of protoplasmic streaming in an intact internodal cell was 78 μ/sec. (Temp.: 28° C.)
(KAMIYA and KURODA 1956 b.)

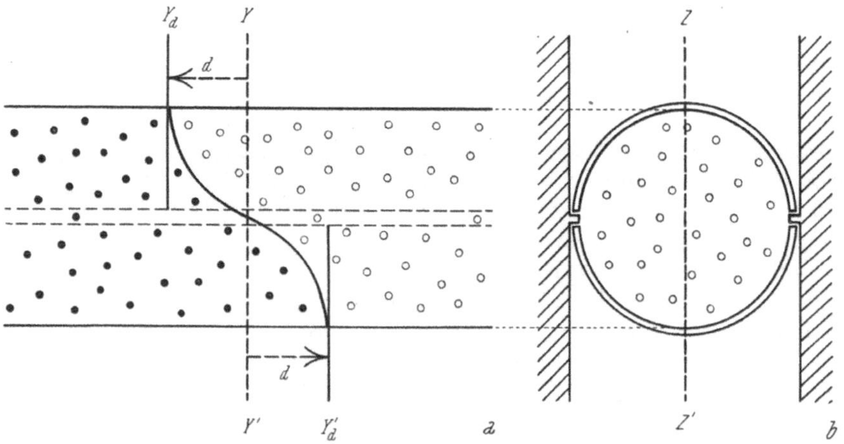

Fig. 69. Model experiment for showing the distribution of a fluid material in a tube when the two equal halves of the tube were made to shift logitudinally against each other. *a* The longitudinal section of the tube after the upper and lower walls were moved from *YY'* line to Y_d and Y'_d respectively. The space represented by solid circles is filled with margarine stained with carmine and that represented by open circles with plain margarine. *b*: The cross section of the tube consisting of two sliding semi-cylindrical walls which are placed face to face in the guiding frame on two projecting rails.
(KAMIYA and KURODA 1956 b.)

intracellular velocity distribution which has the same general pattern as that of the deformation curve shown in Fig. 70, or that of the velocity distribution of the cell sap (Fig. 6), has a significant implication, namely that the endoplasm moves just like an inert fluid finding itself between two semi-cylindrical walls that are made to slide against each other in the longitudinal direction. In case of the living cell, the gel layer itself does not move; nevertheless, we can see from the velocity distribution within the cell that the protoplasmic sol flows as if each half of the gel wall moved in the opposite directions. Thus we have to assume in case of the cell that a force is generated which tends to shift the outermost sol layer to one direction in one longitudinal half of the cell and to the opposite direction in the other half. It is this mechanical energy generated at the interface that becomes the motive force of the protoplasmic streaming occurring in the cell of *Nitella*. All inner portions of endoplasm appear to be moved passively. This kind of analysis indicates all the more clearly that endoplasm alone can take no active part in the streaming. What assumes central importance in the mechanism of the rotational movement in *Nitella* is then the cause of the shifting at the sol-gel interface. KAMIYA and KURODA (1958) calculated that the absolute value of the shifting force exercised at the interface is as low as 1—2 dynes/cm² (p. 40).

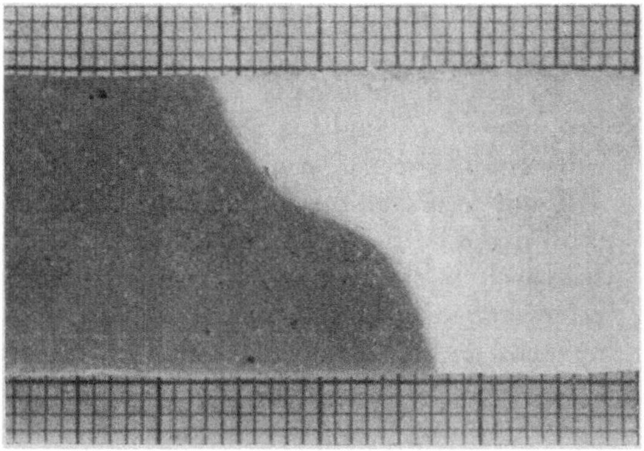

Fig. 70. Deformation of the boundary between red and yellow margarine after the upper and lower walls of the tube (removed before the section was made) were shifted against each other. This shows the longitudinal section through the median plane ZZ' shown in Fig. 69 b. The boundary was originally flat and perpendicular to the longitudinal axis.
(KAMIYA and KURODA 1956 b.)

b) Additional evidences showing the participation of cortical layer in rotation

There are additional facts, if not quantitative, in support of the inference that the organized gel surface is indispensable to the protoplasmic flow.

It has been repeatedly shown that, when the plasmagel layer of *Nitella* or *Chara* cells is injured locally, the endoplasmic flow no longer occurs on the very spot but goes around it (NICHOLS 1925, LINSBAUER 1929, PÉTERFI and YAMAHA 1931). This fact shows that a special organization in the plasmagel layer is essential in maintaining the protoplasmic streaming. Thus LINSBAUER (1929, p. 614) says: "Der Mechanismus der Plasmarotation ist dann nicht im strömenden Plasma selbst oder an der Grenze strömenden Plasma-Zellsaftes zu suchen, setzt vielmehr die Unversehrtheit des peripheren Plasmas voraus. Die Plasmarotation ist mit anderen Worten an die Wechselwirkung von Außen- und Binnenplasma gebunden."

A further fact showing special participation of the cortical gel layer in the protoplasmic flow is presented by BRECKHEIMER-BEYRICH (1954) who stressed in her observation of *Nitella flexilis* under the centrifuge-microscope that the endoplasm accumulated at one end of the cell all in a tumbled mass starts its streaming again along the same course and in the same direction as before the cell was centrifuged. There is no exception to this. Her statement is verified by the author's own experience, as long as the cortical gel remains intact on the wall. In the light of this fact she came to the conclusion that the direction-deciding influence on the protoplasmic streaming does not reside in the endoplasm but is inherent in the inner surface of the cortical plasmagel adhering to the cell wall.

In *Elodea* leaf, too, protoplasmic streaming occurs in the same direction as before in most cells after the leaf has been subjected to ultracentrifugation of as high as 350,000× gravity (BEAMS 1949). In only 7% of the cells could the direction of the streaming be reversed. BEAMS states: "It is difficult to understand where the factors for the polarity of streaming reside within the cell. All of the cytoplasm, including the peripheral membrane, seems to be displaced, at least in so far as could be detected with the microscope." If an analogy to *Nitella* is permitted in *Elodea,* there must nevertheless be a thin ectoplasmic layer left which guides the endoplasm to flow in the original direction. The existence of the membrane, if not visible, must be postulated also for the maintenance of osmotic function.

Another fact relevant to this problem is to be obtained from plasmolysed *Elodea* cells (p. 97) or still better from a naked protoplast taken out of the cell after plasmolysis. We observe there many chloroplasts gathering to form a dish-like clump (systroph) on the bottom of the protoplast. This dish-like aggregate of chloroplasts mounted in cytoplasm rotates as a bloc on an axis without very much changing their relative position. Sliding occurs in this case at the inner surface of the ectoplast which is very thin. As the angular velocity is not very different at the central and peripheral regions of the aggregate, the actual rate of shift relative to the ectoplast is the greater the farther from the center of rotation.

Generally, in a cell conducting typical rotation, such as in the case of *Nitella* or *Elodea,* the facts obtained from different angles are focused upon the same point—the essential role played by the sol-gel interface in the occurrence of the streaming.

2. Streaming caused passively by difference in internal pressure

The above examples represent cases in which the seat of the motive force appears to reside at the interfacial region between quiescent endoplasmic gel and moving endoplasmic sol. There are, however, cases in which the mode of streaming is apparently different.

As already referred to (p. 21), the velocity distribution of the plasmodial strand of the myxomycete, *Physarum polycephalum,* takes the shape of a parabola the apex of which is flattened or truncate. On the basis of analysis of this shape, KAMIYA (1950 a) came to the conclusion that the intracapillary flow of protoplasm is caused passively by a pressure difference somehow established between the anterior and posterior parts of the flow.

This conclusion was corroborated by a further experiment (KAMIYA and KURODA 1958 a) which makes possible a comparison of the velocity distribution of the protoplasmic streaming in one and the same plasmodial strand under two different conditions, *i. e.*, first, the streaming under the normal state of the plasmodium and, second, streaming induced artificially by using the double-chamber method (p. 40). They obtained the artificially induced passive flow in the plasmodial strand by applying appropriate air pressure to one of the two compartments at the very moment in which the direction of natural flow reverses and consequently the streaming stops momentarily. The velocity distributions were obtained cinematographically.

At the left in Fig. 7 (p. 21) is shown a microscopic view of a part of the connecting strand consisting of the capillary tube. Fig. 7 *a* shows the velocities at a time when protoplasmic streaming occurs spontaneously in it while Fig. 7 *b* shows the velocity distribution of the flow caused by a pressure of 15 mm. of water column. Both *a* and *b* show the distances covered by small particles during 3 seconds in one and the same portion of the same connecting strand, the length and bore of which are 5 mm. and about 0.1 mm., respectively. The only difference in condition between the two is that the flow in *a* is spontaneous whereas that in *b* is passively induced. As the rate of the natural flow is ever changing, we looked for a definite short interval in which the rate of the flow is equal to that of the flow artificially induced so that the comparison can be made exactly. Thus by superposing the velocity distribution of the artificial flow on that of the natural flow we can realize that the forms of the two are identical.

If the driving mechanism would reside, for instance, at the boundary between the plasmagel (wall of the capillary tube) and plasmasol (endoplasm) as in the case of the *Nitella* cell, the shape of the velocity distribution would be different between natural streaming and that caused artificially by pressure applied from without. This is a cogent evidence that protoplasmic streaming in the capillary tube is brought about by a difference in the internal pressure of the plasmodium. In other words, the endoplasmic streaming in the capillary tube is a passive phenomenon.

It is well known that when a true viscous fluid flows through a circular tube in the laminar flow state, the particles of any transverse section at a given moment will, after the lapse of a certain time, be found on the surface of the paraboloid; this fact is a necessary consequence of the POISEUILLE-HAGEN's law. It is known, on the other hand, that the figure of extrusion of the liquids having structural viscosity shows usually a curve like a parabola the apex of which is flattened. The width of the flattened part of the anomalous flow is naturally dependent upon the shearing stress as well as on the rheological properties of the fluid. If the fluid has a yield value, the apex of the parabola is truncate. Anyway, the flattened part becomes wider as the shearing stress gets smaller and hence the flow slows down. In other words, when the driving force (pressure difference) increases, the intracapillary velocity distribution approaches the revolution of a perfect parabola.

Thus, there is sufficient experimental ground to conclude that the intracapillary flow of the slime mould is induced passively by a difference in local pressure induced somehow inside a plasmodium. The problem of the mechanism of the protoplasmic streaming in the slime mould is, therefore, reduced to the question how a local difference in pressure is established in a protoplasmic system.

The so-called tidal streaming of protoplasm occurring in hyphae of fungi [*Mucor, Rhizopus, Phycomyces* (Arthur 1897, Schröter 1905, Andrews 1912, Grehn 1932, etc.); *Ascophanus* (Ternetz 1900); *Humaria* (Jahn 1934) etc. cf. p. 11.] has a characteristic in common with that in a plasmodial strand in that the flow is caused passively by a local difference in pressure within a continuous hyphal system. In the tidal streaming, however, the establishment of a pressure difference is closely related with osmotic phenomena, which is not the case in the myxomycete plasmodium. The sleeve-type streaming in coenocytic hyphae is another matter.

3. Protoplasmic motion in an isolated protoplasmic fragment

An effective method to investigate the mechanism of the streaming is to take the protoplasm out of the cell by operation and to study in detail the behaviour of these isolated droplets. The fragment of protoplasm in various plant cells was studied long years ago by many workers each for a purpose of his own—Hanstein (1880), Sachs (1892), Nichols (1922) in *Vaucheria,* Pfeffer (1877) in *Hydrocharis,* Ewart (1903) and Yotsuyanagi (1953 a) in *Elodea,* Küster in *Tradescantia* (1933 b) and *Bryopsis* (1939), Weide (1939) and Schorr (1939) in *Phycomyces,* Iwanami (1952, 1956) in *Lilium* (pollen tube), etc. Besides these, the protoplasmic fragment obtained from cells of *Characeae* was especially well studied with respect to motion.

That chloroplasts in an isolated drop of protoplasm rotate on their own axes was known by Donné (1838) and Dutrochet and Brongniart (1838) 120 years ago. Later the phenomenon was repeatedly noticed and confirmed by many authors—Velten (1876 b), Hörmann (1898), Stålberg (1927), Valkanov (1934), Yotsuyanagi (1953 a), Jarosch (1956 a), Kamiya and Kuroda (1957 b) etc. Rotation of nuclei was observed by Valkanov (1934) Yotsuyanagi (1953 a) and recently by Fetzmann (1957, 1958). As to the mass streaming of the protoplasm in the isolated drop, it was observed by Jost (1929) and particulary by Yotsuyanagi (1953 a, b). The latter author further noticed various types of motion including rhythmic pulsation (p. 62), amoeboid movement, sliding movement (p. 149) etc. of the protoplasmic fragments isolated from *Chara* and *Nitella* cells. On the other hand, Linsbauer (1933) got negative results in trying to find an active motion of or in protoplasmic drops of *Nitella.*

The intriguing motions, which were adequately described by Yotsuyanagi, can be verified by squeezing out both endoplasm and ectoplasm forcefully from the cell. It is to be noticed in this case that drops in one and the same preparation exhibit not just the same type of motion but several different types all at the same time. The question naturally arises as to why it is that different fragments of protoplasm coming out of one and the same cell into one and the same milieu behave differently. What seems to be important in this question is the fact that the drops of protoplasm obtained by YOTSUYANAGI and many other authors are not the same in the proportions of the cortical plasm and the endoplasm composing them. As has been pointed out by YOTSUYANAGI, there is a certain concentration range of solutes [$CaCl_2$ or $Ca(NO_3)_2$, KNO_3, glucose] in which a certain type of movement can take place, but the variability of the motion in one and the same medium is most probably to be ascribed to the differences in the relative amount and the mode of existence of the cortical gel in the squeezed-out fragments. Discrepancies among the observations of previous authors are probably due at least in part to these differences.

Hence, in order to analyse the dynamic functions of the endoplasm and the cortical plasm, it seems important as basic information to know how an endoplasmic drop containing no cortical plasm behaves. For isolating a naked drop of protoplasm, most of the previous authors amputated a cell and squeezed out the contents keeping the protoplasm either in the natural cell sap or in appropriate artificial media; they seem to have taken no special precaution to separate the endoplasm from the cortical plasm.

a) Effusion of endoplasm

Generally, when an internodal cell of *Nitella* is cut into two pieces, these pieces are crushed as the turgidity is lost. The cortical gel is detached from the cell wall together with the imbedded chloroplasts and a part of it is carried away with the endoplasm to the outside of the cell. In order to protect the cell as far as possible from mechanical damage caused by amputation, KAMIYA and KURODA (1957 a) used a suitable pressure-chamber. An internodal cell is placed in a vertical position across the glass wall of the chamber through a narrow groove in such a way that a greater part of the cell is inside and only a small portion of its lower end is left outside. The portion of the cell sticking out of the chamber is submerged in an appropriate solution [0.08 M KNO_3 + 0.05 M $NaCl$ + 0.004 M $Ca(NO_3)_2$] in a small observation vessel. The pressure-chamber is also filled with the same solution.

While applying a weak negative hydrostatic pressure amounting to about 2 cm. of mercury inside the chamber, amputation was conducted with sharp scissors at the lower portion of the cell which is outside the pressure-chamber but is submerged in the solution in the vessel. In virtue of the negative hydrostatic pressure the part of the cell placed in the chamber keeps its turgidity even when it is cut at one end and left open. When seen through the horizontal microscope, it is noted that the cortical layer of the cell part inside the pressure-chamber suffers no visible alterations and the endoplasm continues to flow along the cortical gel layer with the same rate as that before amputation, as if it were not aware of this seemingly drastic operation. Injury is limited to the direct proximity of the locus where amputation was conducted. When the endoplasm creeping along the cortical gel reaches the opening, it comes out of the cell and falls down gradually and smoothly into the medium in the observation vessel as a continuous thread (Fig. 71).

The thread of endoplasm does not disperse into the solution but falls down with clear boundaries onto the bottom of the vessel where it forms a sessile drop. The effusion of protoplasm occurs only on one side of the cell opening where the downward streaming comes across. When the amount of endoplasm coming out of the cell and that going into the drop on the bottom are equal, the phenomenon is stationary; it appears in this case that the opening of the cell and the sessile drop on the bottom are connected with an endoplasmic column, as it were. The effusion lasts 20–40 minutes without interruption, the duration depending upon the size of the cell. This duration approximately corresponds to the time required for the endoplasm, which had passed beyond the site of amputation a little before it was performed, to reach the opposite side of the same opening after travelling all the way around the inside of the cell. When the supply of endoplasm becomes scanty, the continuous thread of protoplasm coming out of the cell becomes intermittent, but the effusion of the endoplasm still continues for some time by repeatedly forming a pendant drop until the endoplasm is nearly exhausted. A series of pictures shown in Fig 72 *a–e* shows successively how a pendant drop of endoplasm is formed, becomes larger and changes its shape until eventually it falls onto the bottom where it forms a sessile drop.

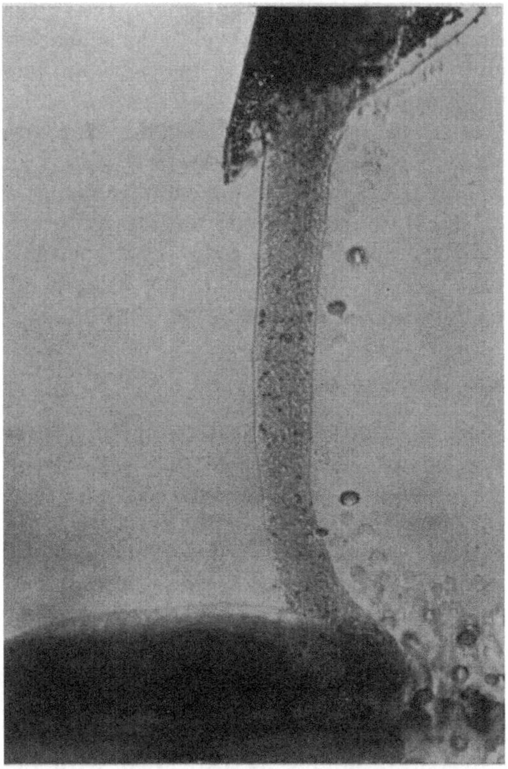

Fig. 71. Effusion of the endoplasm in the form of a continuous thread from the lower opening of the internodal cell of *Nitella flexilis*. On the bottom of the cuvette is seen a large drop of the naked endoplasm. (Taken about 30 minutes after amputation.)
(KAMIYA and KURODA 1957 a.)

As for the cortical gel in the cell in which most of the chloroplasts are imbedded, it remains intact showing no indication of injury whatsoever. This fact gives us ground to assume that the protoplasm coming out of the cell in this manner is no other than the endoplasm which has been taking part in the protoplasmic streaming when it was in the cell.

As mentioned in the foregoing, analysis of intracellular velocity distribution has led us to conclude that the endoplasm alone must be inert in respect to motion. From this viewpoint special attention is to be paid now to its behaviour when it comes out of the cell. It is clearly noted that the endoplasm falls down gently as a whole like a plug without giving rise to any appreciable streaming inside. The only visible movement in the falling column of endoplasm is the Brownian motion of the small particles and the independent rotation of chloroplasts on their own axes. That the endoplasm is poor in elasticity was also demonstrated by the fact that the endoplasmic column bends like a loop worm rather than contracts when its two ends are brought closer (KAMIYA and KURODA 1957 b).

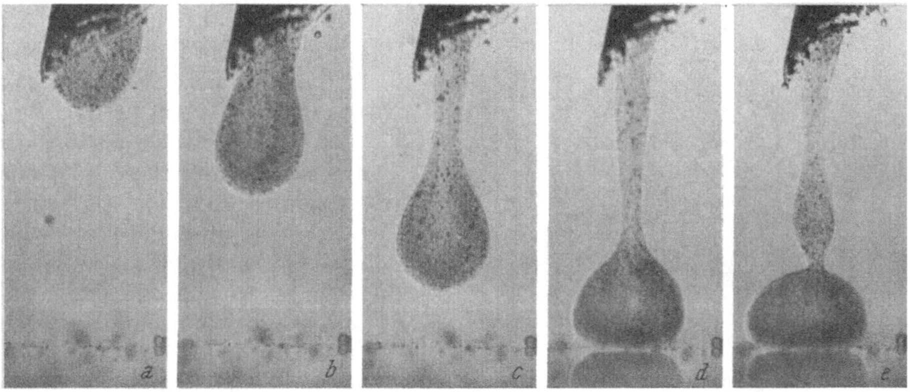

Fig. 72. Successive stages showing the formation of a pendant drop of the endoplasm at the opening of the *Nitella* cell and its subsequent transformation to the sessile drop on the bottom of the cuvette. The cell is the same as that in Fig. 71. *a*: about 90 min. after amputation, *b*: 90 sec. after *a*, *c*: 15 sec. after *b*, *d*: 10 sec. after *c*, *e*: 63 sec. after *d*.
(KAMIYA and KURODA 1957 a.)

b) Protoplasmic motion in an endoplasmic drop

The endoplasm having reached the bottom of the cuvette is divided later into several drops with diameters of 150–350 μ whose profiles are like those of mercury drops of various sizes and enable us to calculate the tension at the surface of the naked endoplasm. It is as low as 0.001 to 0.006 dynes/cm. (KAMIYA and KURODA 1958 c). These endoplasmic drops seem to suffer least injury as compared with the protoplasmic drops so far obtained by previous authors and the former are further different in that they consist entirely of endoplasm while the latter involve more or less the cortical gel which was exfoliated by mechanical force. As the endoplasm of a normal internodal cell of *Nitella* contains nuclei and also some chloroplasts, so do the endoplasmic drops thus obtained.

Both nuclei and chloroplasts being somewhat greater in their specific gravity than their milieu (ground plasm, or hyaloplasm) they tend to sink in the lower part of the drop. What we are concerned with is the motion in the drops of endoplasm on the bottom of the cuvette which can be observed from below by means of an inverted microscope while the drops are left as they are (Fig. 73).

The point to be stressed here is that there is no mass streaming in the isolated drops of the endoplasm thus obtained, yet chloroplasts as well as nuclei rotate most actively each on its own axis. This fact is a re-confirmation of the observations reported by some of the previous authors, especially by VALKANOV (1934) on nuclei and by JAROSCH (1956 a) on chloroplasts. Chloroplasts found in an isolated drop of endoplasm have been suspendend in the endoplasm before the cell was cut and have not been stripped off from the cortical gel layer as a result of operation. The rate and direction of rotation vary with individual chloroplasts, the rate being generally within the range of 0.5–3 revolutions per second. The rotation of the chloroplasts, which can serve as an indication of the survival of the drop, lasts usually for 10–50 hours.

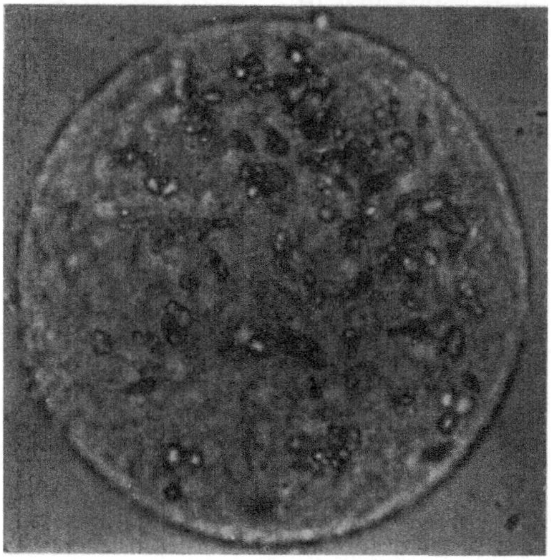

Fig. 73. An endoplasmic drop (*Nitella flexilis*) sunk to the bottom of a cuvette seen from below. Diameter of the drop: 210 μ. (Reproduced from a 16 mm. cinematographic film.)
(KAMIYA and KURODA 1957 b.)

An important question now arises as to why there is no mass streaming in the drops consisting solely of endoplasm which was in a state of active motion when it was in the cell, while individual chloroplasts rotate furiously on their axes.

Since the drops are lacking in cortical plasm, absence of mass streaming is rather an evidence in support of the view arrived at in the foregoing that the endoplasmic sol is moved passively and that it alone is not capable of producing the flow. Just as the streaming in an intact cell is caused by the shifting force generated at the interface between sol (endoplasm) and gel (cortical gel), we think that the free rotation of the chloroplasts is to be attributed to a shifting force generated at the interface between sol (endoplasm) and gel (chloroplasts or nuclei). Probably a particular gel structure, which is functionally comparable with or identical to that of the cortical gel, is present at the surface of the freely suspended particles (chloroplasts or nuclei). Such a view was held already by HÖRMANN (1898) and recently by JAROSCH (1956 a). Under such circumstances, it is not difficult to understand that when the gel phase is fixed as in the case of the normal cell, streaming of the sol phase (endoplasmic streaming) takes place, and that when the gel phase is freely suspended, as in the case of an endoplasmic drop without cortical gel, the gel phase itself moves with a force of the same nature. If the impulse-producing mechanism exists only partially on the surface of the particles, they tend to rotate rather than travel around. But if the chloroplasts come out of the cell tied in a row as was the case when they were in the cortical layer, they move around in the endoplasm in the longitudinal direction of the row without individual rotation. Thus free motion of the gel particles (chloroplasts and nuclei) in the isolated endoplasmic drop is quite in conformity with the conclusion that the seat of the force is confined to the interface region between the plasmasol and the organized plasmagel.

The phenomenon which shows the relation of a reciprocal shift still more clearly is a minute, local streamlet observable in the region very close to the surface of the gel particles, whether nuclei or chloroplasts. The direction of the streamlet is always opposite to that of the rotation of the particle. This minute, but rapid stream of plasmasol in the direct proximity of a nucleus or chloroplast was noticed a long time ago by several workers (Donné 1838, Dutrochet and Brogniart 1838, Velten 1876 b, Hörmann 1898,

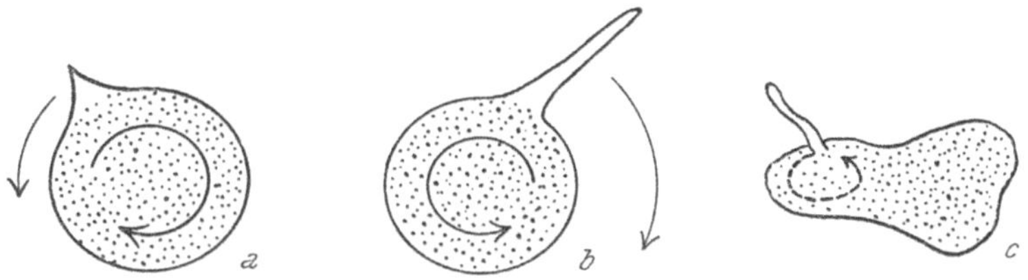

Fig. 74. Characteristic movements displayed by some of the isolated drops of protoplasm (*Characeae*). (*a*) A small pointed projection turns around on the protoplasmic drop. At the same time the interior protoplasm rotates in the direction opposite to the movement of the projection. (*b*) A digitate protuberance turns around the protoplasmic drop with simultaneous rotation of the interior protoplasm in the opposite direction. (*c*) A protuberance is displaced along the circular track on the surface of an amorphous piece of protoplasm. (Yotsuyanagi 1953 b.)

Stålberg 1927, etc.) stressed by Valkanov (1934), and has been studied most exactly by Jarosch (1956 a) recently. Observations of these authors are unanimous in that there is no mass streaming nor change in the contour of the protoplasmic drop and in that there is vigorous independent motion of chloroplasts or nuclei with simultaneous streamlets occurring in the reverse direction. Probably the objects of observation of these authors were the drops containing only a small fraction of cortical plasm, if any at all.

On the other hand, many types of motion other than the above have been described by Yotsuyanagi in the naked protoplasmic fragments of *Characeae*. Of these, the development of orderliness from agitation to rotation (p. 16), rhythmic contraction and undulation (p. 62) has been already mentioned. Of special interest for the problem in question is the sliding motion noted by him.

c) Sliding motion of protuberances at the protoplasmic surface

Among many protoplasmic fragments squeezed out mechanically from *Nitella* or Chara cells by Yotsuyanagi, small drops (5–30 μ in diameter) are found on which one or many little protruding parts are formed. These protruding parts are entirely hyaline and are sometimes possessed each of a sharp point-like little horn (Fig. 74 *a*), sometimes more elongated in the form of a digitate protuberance with a sharp or rounded points (Fig. 74 *b*). It is remarkable enough that they turn around the drop sliding on the surface. This sliding movement is always accompanied with another movement in the interior, a rotational movement of the entire content of the protoplasmic drop in the direction opposite to that of the movement of protuberances. No exception has been found to these two facts; that is, that the two movements above take place always together, and that their directions are always opposite. This situation is comparable with that of the free rotation of chloroplasts described in the foregoing in that the two movements are involved in it, *i. e.*, rotation of a chloroplast or nucleus itself and the concurrent counter-streaming of the surrounding medium. Though it is difficult to analyse how such a motion can be brought about, it appears that rigid protuberances move by sliding with their feet on the protoplasmic surface. In a special case, a protuberance is produced on the surface of the protoplasmic drop that is entirely amorphous, and slides regularly through a circular track situated somewhere on the surface (Fig. 74 *c*). All these rather curious behaviours of the protoplasmic fragments are, in the author's opinion, disclosing in a characteristic manner a phenomenon identical to that of the shifting occurring at the interface between cortical gel and endoplasm. Such a phenomenon is never observed in a drop consisting solely of the endoplasm. The hyaline protuberance is probably made of the cortical plasm of the cell, but it does not surround the drop nor is held by the cell wall; it is adhering to the endoplasmic drop in a bizarre manner only over a small area. Hence, if there is a mechanism causing an impulse between the foot of the protuberance and the endoplasm, the hyaline gel (protuberance) itself would move with the concurring counter-rotation of the endoplasm. Strange though the motion may seem, this is again a phenomenon of interfacial shifting occurring in a small fragment of a protoplasmic system consisting of both endoplasm and ectoplasm.

4. Moving fibrils in protoplasm

Though little is known about the nature of the force exercised at the interface between the cortical gel and endoplasm, a precise observation made by Jarosch (1956 a, 1957) contributes to this problem not a little from the morphological side. Just as the previous authors did, Jarosch cut an internodal cell of *Chara foetida* and squeezed out the protoplasmic drops in the natural vacuolar sap. He recognized in an isolated protoplasmic drop the following components: nuclei, chloroplasts, protein lumps in irregular forms, spherosomes, extremely tiny particles, and fibrous bodies of more variable occurrence. The last components are often so very thin and engaged in such an active motion that only a well-trained eye can notice them by means of dark field illumination and an oil immersion lens (Fig. 75). Photographs in Fig 76 *a, b* show these fibrous bodies observed in the author's laboratory in protoplasmic drops obtained from internodal cells of *Nitella flexilis*. The drops were squeezed out together with the cell sap and cortical plasm. The behaviour of these fibrous bodies indicates their significant bearing on the protoplasmic movement.

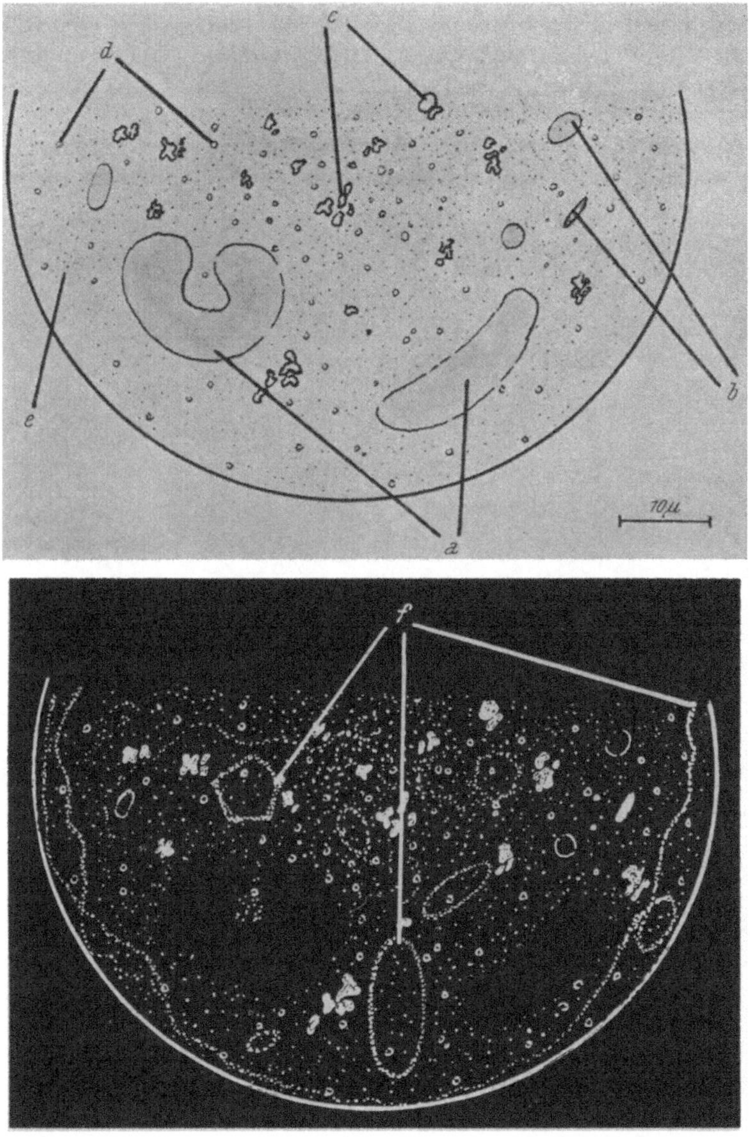

Fig. 75. A part of protoplasmic drop squeezed out from *Chara foetida* in bright (top) and dark (bottom) field. *a*: cell nucleus, *b*: rotating chloroplasts, *c*: protein lumps, *d*: spherosomes in Brownian movement or agitation, *e*: small microsomes, *f*: moving fibrils, in which some small granules are enclosed; these fibrils are invisible in bright field.
(JAROSCH 1956 b.)

Actually, these fibrous bodies are bundles of still finer filaments on the submicroscopic level and move in their longitudinal directions independently. Though variable in occurrence, they often form closed rings, pentagons, or hexagons. These closed rings rotate sometimes so rapidly that the eye cannot follow their movement; they give a flimmering appearance. The rate of revolution of the ring which can be observed by the microsomes adhering to it is estimated to be 4–5 rev./sec. at the maximum and 1–2 rev./sec. on the average. The moving strands and rings contain mostly microsomes, but the strands and rings themselves also gleam faintly. It is thus also possible to recognize microsome-free rings.

What is characteristic of the threads or rings besides their motility is their instability. The same ring is not to be observed for more than

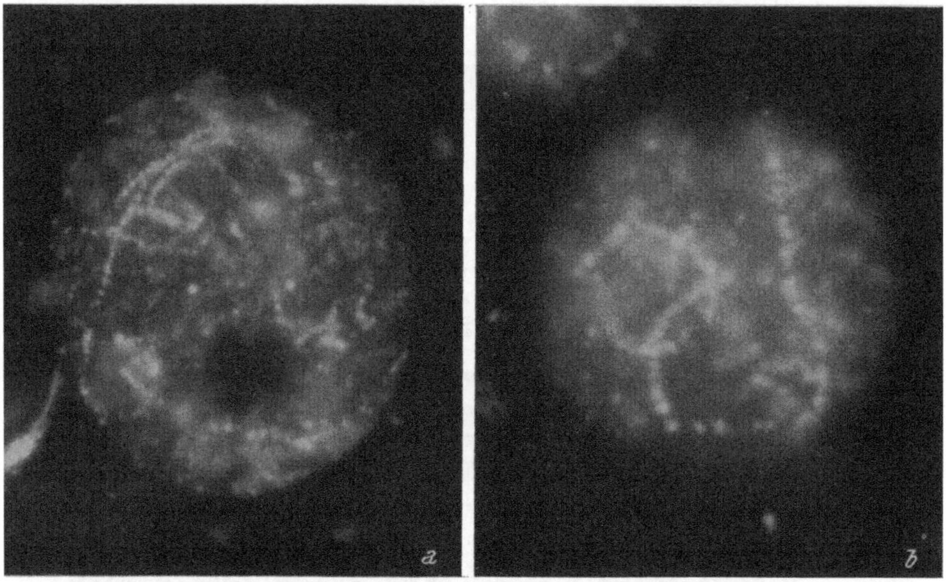

Fig. 76. Moving fibrils carrying microsomes in isolated drops of protoplasm (*Nitella flexilis*). *a*: major axis of the drop: 55 μ, *b*: diameter of the drop: 25 μ. (Dark field illumination.) (Original.)

20–30 minutes. A conspicuous change in thickness noticed in the same ring is likely to mean the "roll in" and "pay out" of still finer, submicroscopic fibrils. This rolling-in and paying-out process must occur one after the other without the direction of rotation of the ring being altered. When the reeling-up predominates, the ring gets thicker; when the reeling-off predominates, the ring becomes thinner. Not seldom disintegration of the ring into thinner threads can be observed directly. Fig. 77 shows this. The ring exhibiting gentle rotation breaks down into a spiral-like thread. It is probably made of continuous thread which may be split up further, eventually disappearing beyond the limit of microscopic resolution. These observations definitely show that a visible thread is not an elementary one but is composed of thinner threads. Integration and disintegration of fibrous bodies probably proceed simultaneously in one and the same isolated drop of protoplasm, since threads and rings which have been clearly visible disappear in the course of time and new ones emerge at other loci. When the ring breaks up into finer fibrils, these fibrils move faster than the ring which they have built up before.

Very interesting is a regular undulating motion displayed by the thicker threads or rings. The bending of a thread or ring is transmitted as a wave in the same direction as that of the bodily motion of the thread or the ring, but the former goes somewhat faster than the latter. Sometimes a linear stretch of a constant length moves around along the ring (cf. Fig. 76 *b*, bottom of the protoplasmic drop). This form is often transferred to the regular pentagons or hexagons. The undulation is seen only in thick threads. In case of the thick ring, undulating motion is vigorous, though its rotation is insignificant. The fibrils, split from the ring, show a very active forward motion but not a trace of undulation. The inference may be made then that a thicker thread, which can perform wave motion, but which scarcely moves forward, is composed of thinner fibrils, which are able only to glide forward. The undulating motion of such a protoplasmic thread reminds us of a slow flagellar motion. Though contorted or vibrating motion of fibrils in the protoplasm has been noticed by some early workers under various conditions (Brücke 1862, Schultze 1863, Luerssen 1868), it is Jarosch who noticed and stressed the regularity of the motion. Scarth (1927, 1942) discussed Strasburger's kinoplasm as a basic differentiation of the cytoplasm and pointed out its possible role in the protoplasmic streaming of a plant cell. The fibrous bodies described by Jarosch seem to be closely related to the kinoplasm, but for a decisive answer as to whether the former is identical with the latter which is known to be rich in lipoid, further study is needed.

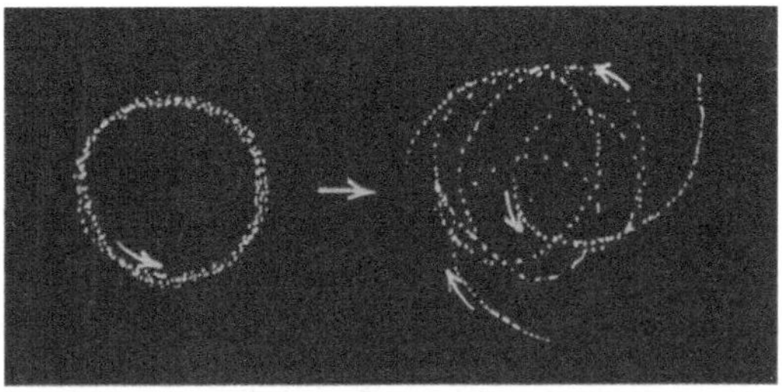

Fig. 77. A thicker ring rotating slowly in the counter-clockwise direction disintegrates in a few seconds into thinner threads which are barely recognizable. These thinner threads move quicker in both clockwise and counter-clockwise directions.
(Jarosch 1956 b.)

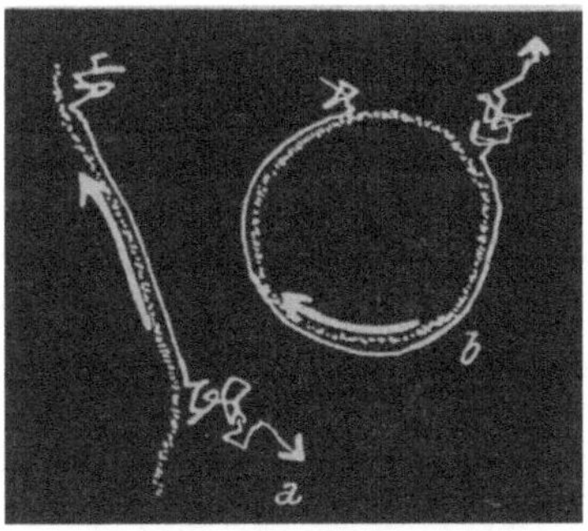

Fig. 78. Behaviour of microsomes in the proximity of the moving fibrils. *a*: a microsome which has shown only Brownian movement at first, receives a countermoving impulse in the proximity of the creeping thread. *b*: the same in the proximity of the rotating ring.
(Jarosch 1956 b.)

An important point in respect to Jarosch's observation is the mechanism of the motion of the thread, especially of that enclosed in a ring. Naturally, no concrete mechanism of it is yet known, but there is an observation which helps throw some light upon this problem. It is a motion of microsomes along the fibril occurring in the direction opposite to the movement of the fibril itself. The relation of the movement of microsomes to that of the protoplasmic fibril is just the same as the relation of the movement of microsomes to the free rotation of a chloroplast or of a nucleus. While a microsome receives the counter-impulse from the fibril, no Brownian motion is to be seen, as the particle is stabilized in one way or another, but the particle exhibits Brownian motion again as soon as it leaves the action sphere of the fibril. Fig. 78 *a* represents a creeping thread. When a microsome, which showed Brownian motion at first, comes in the action sphere of the thread, it suddenly moves along it in the direction opposite to that of the movement of the thread. A similar phenomenon is exhibited also in the case of a rotating ring (Fig. 78 *b*).

In view of the fact that fibrous elements are present which are capable of a definitely directed movement and which often form rotating rings, JAROSCH assumed that the rotation of the chloroplasts is due to the adherence of these threads to the chloroplasts. As a matter of fact it can sometimes be directly recognized under the ultramicroscope that these threads are attached to the chloroplasts. The impulse relationship between the threads and the counter-streaming of the medium surrounding them is comparable to the relation between a rowboat and the water that is pushed back by the oars. These fibrous elements are also believed to be arranged with regular orientation on the inner surface of the cortical gel. The normal protoplasmic streaming is then explained as being due to the impulse action of these fibrils. This interpretation of JAROSCH is not incompatible with many observations made so far.

Thus it is possible to hypothesize that streaming in *Nitella* is brought about by an impulse of unknown physical nature from the fibrils which are possibly arranged in a regular orientation on the inner surface of the cortical layer. If, from a cortical layer, a chloroplast is detached while these impulse-inducing fibrils remain attached to it, then it must move in the same way as a rowboat, which when anchored produces only a stream of water. This is the outline of JAROSCH's hypothesis.

YOTSUYANAGI (1953 b) observed protoplasmic fragments which disaggregate in a very peculiar manner in a certain concentration range of the external milieu [0.24–0.25 M $Ca(NO_3)_2$ or $CaCl_2$]. A great number of thin protuberances resembling myelin figures come out of the surface of the protoplasmic drop. The myelin-like filaments can also be formed toward the interior of the drop instead of projecting toward the outside. In this case the concentric zones with a strong refractive index are produced in the form of fingerprints which are clearly separated from the granule-rich part of the cytoplasm.

On the other hand, JAROSCH noticed a similar structure in old preparations when protoplasm degenerates. Such a fingerprint-like structure changes its form gradually but steadily. The Brownian movement and protoplasmic movement are directed along it. Obviously the structures observed by YOTSUYANAGI and JAROSCH are the same thing.

Of special interest also is the movement of protoplasmic pieces, which take, strangely enough, geometrically regular forms, such as ellipses, equilateral triangles, regular hexagons, etc. All the points on the contour of the geometrical figure are shifted in the same direction with the same speed. Consequently, we see there a geometrical figure rotate apparently as if it were rotating against the substrate. Exact observation reveals, however, that it is again a matter of undulation on the membrane.

Some of the data obtained by YOTSUYANAGI show: 1. an ellipse, with a great axis of 7 μ, a small axis of 5 μ, a period of rotation of 35.5 seconds; 2. an equilateral triangle, with a side of 18 μ, which kept this regular form during 20 minutes in the course of its rotatory movement, then began to take diverse forms of irregular triangles one after another, the rotatory movement becoming less active, until at last it lost the triangular form when the movement ended; 3. regular hexagons, with a side of 5 μ, which continued the rotation during at least 5 minutes in this form, and came back at last to a spherical form when the movement ended. YOTSUYANAGI observed also many irregular polygons whose angles were so numerous that it was impossible to count them in the moving state.

Although JAROSCH conducted his interesting experiments probably without being aware of YOTSUYANAGI's work, it is interesting to compare some of the data obtained by these two authors. Especially the fact that YOTSUYANAGI observed small protoplasmic pieces whose contour takes the form of isolateral polygons and that JAROSCH observed isolateral polygons formed by protoplasmic fibrils in a large protoplasmic drop, leads us to assume some interrelation between the two. Since the modes of movement of these geometrical structures and their dimensions are also of the same order, it is possible to assume that YOTSUYANAGI's polygons or ellipses are no other than small protoplasmic pieces tightly occupied by JAROSCH's polygon or ring. It is probable that variability in the type of motion displayed by the isolated fragment of protoplasm reported by YOTSUYANAGI (1953 a, b) is due to the amount and mode of action of JAROSCH's fibrils. If these fibrils form thicker bundles or the cortical layer of the naked drop, the isolated fragment may show a rhythmic undulation of its contour (Fig. 36) or an amoeboid motion, which are also observed by YOTSUYANAGI.

Thus it is possible to hypothesize that streaming in *Nitella* is brought about by an impulse of unknown physical nature from the fibrils which are possibly arranged in a regular orientation on the inner surface of the cortical layer. We must, however, naturally be very cautious in generalizing the ideas gained in cells of *Chara* or *Nitella*, which show a steady and hardly variable orientation of the fibrils in the cortical structure. In other rotations, especially in the induced ones such as in the *Elodea* cell, the position of the fibrils must be more easily changeable. In the circulation type of streaming the tie is perhaps still smaller so that most of the fibril complexes can fall apart from the wall layer and build the strings.

5. The phenomenon of shifting occurring in the plasmagel

Generally speaking, there is in gel structures some connections among the micells that form them. In inanimate gels these connections are permanent in most of the cases, so that the micells do not repeatedly connect and disconnect themselves under a definite condition. In living protoplasm, however, as Frey-Wyssling (1953, p. 178) adequately pointed out, even when it is apparently in a gel state, the structural elements repeatedly join together and fall apart. Moreover, there are cases when a new junction occurring after the disappearance of another should be considered as being produced by the shifting of the previous junction toward a certain direction. That is to say we see a tendency in certain plasmagels to form a shear by itself. Since this tendency seems to have its basic character related to the shifting occurring at the sol-gel interface in the case of protoplasmic streaming, the phenomenon is dealt with here. A good example of it is seen in the slime mould.

It has been found out by Kamiya and Seifriz (1954) that when the filamentous part of a slime mould (*Physarum polycephalum*) is taken out and suspended in moist air by supporting its two ends with two protoplasmic masses, it presents first a catenary form, then gradually becomes spiral, finally to form loops and twists. These look very much like the loops and twists that are formed when twisting a string at its both ends. Such a phenomenon suggests that a force bringing about shear within the structure has been generated spontaneously in the protoplasmic string.

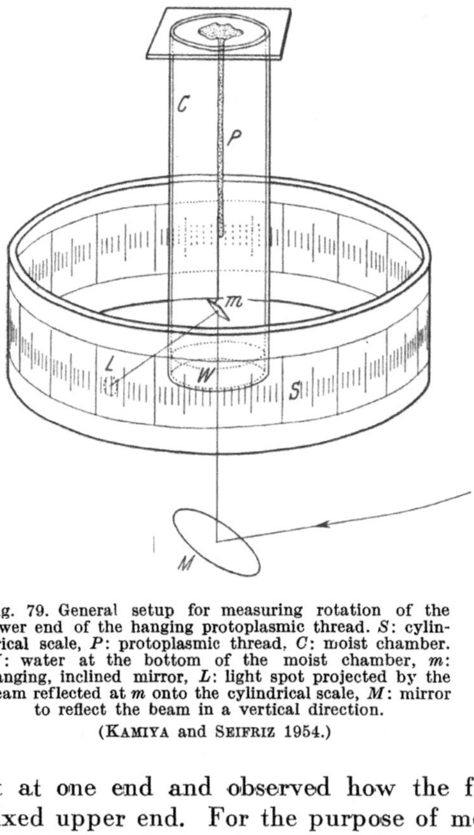

Fig. 79. General setup for measuring rotation of the lower end of the hanging protoplasmic thread. *S*: cylindrical scale, *P*: protoplasmic thread, *C*: moist chamber. *W*: water at the bottom of the moist chamber, *m*: hanging, inclined mirror, *L*: light spot projected by the beam reflected at *m* onto the cylindrical scale, *M*: mirror to reflect the beam in a vertical direction. (Kamiya and Seifriz 1954.)

In order to measure this phenomenon quantitatively, Kamiya and Seifriz suspended the protoplasmic string perpendicularly in a moist chamber by fixing it at one end and observed how the free lower end twists against the fixed upper end. For the purpose of measuring exactly the angles of the twisting, a minute mirror slanted at an angle of 45⁰ against the horizontal surface was fixed at the lower end of the protoplasmic string, so as to have it reflect horizontally the light coming from below and project it on the cylindrical scale set up around it (Fig. 79).

Fig. 80 shows an example of the actual measurement of the twisting of a protoplasmic string. As the figure has been drawn in such a way that (+) shows the case when the lower end of the string twists counterclockwise as seen from above and (—) when it twists clockwise, the gradual rise of the wave-like curve delineated with open circles at each wave means that the protoplasmic string, repeatedly twisting itself to the right and the left, has twisted more in the counter-clockwise direction at each twist. In Fig. 80, we see that the protoplasmic string has twisted itself irreversibly in the (+) direction about 16 times in 20 minutes. Such a movement continues as long as the protoplasm is living and ceases with its death. Twisting in the (—) direction, that is clockwise, is

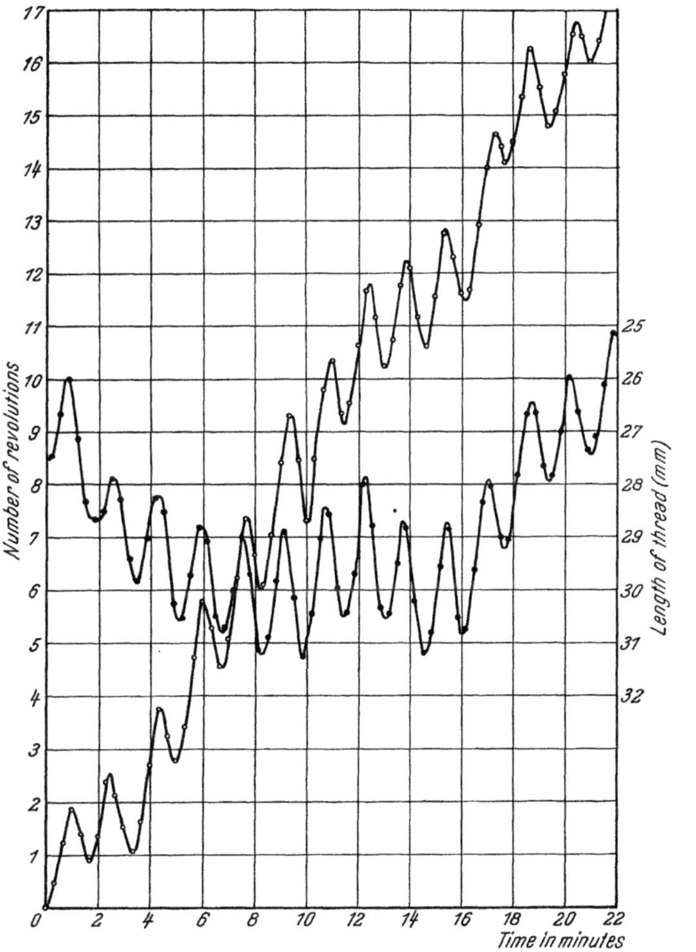

Fig. 80. Time course of rotation (open circles) of the free end of the thread and that of elongation and contraction of the thread (solid circles).
(KAMIYA and SEIFRIZ 1954.)

also seen in certain materials, but statistically, counter-clockwise twisting is more frequently seen. In 52 cases, 33 showed a (+) rotation and 19 a (—) rotation. Obviously, the predominance of (+) rotation over (—) rotation is not due to a mere chance, but the reason for it is unknown. As is shown by the undulating curve drawn with solid circles in Fig. 80, the hanging thread elongates and contracts periodically by about 10% of its entire length at the same time as it twists. Its periods coincide with the periods of twisting, but the relation of the phases of the two waves is not necessarily the same in all materials. It was also shown clearly that the protoplasmic streaming which is sometimes observable in the hanging thread has nothing to do with the cause of the torsion, although coincidence of the period of streaming and the period of torsion suggests that both types of motion are controlled by one and the same basic physiological rhythm.

When the temperature is raised from 24° C. to 30° C. during the experiment, not only the angular velocity of the rotation increases, but also the oscillation in the counter-direction apparently disappears and the lower end rotates continuously only in the (+) direction. It was also observed that when the temperature is lowered to the normal, the rotation returns to the former mode of movement (KAMIYA and SAITO, unpublished).

These results are noteworthy in connection with the dynamic nature of protoplasm. The protoplasmic string used in the experiment is usually 30–150 μ in diameter and 20–60 mm. in length, but as it supports the weight of the mirror (about 2 mg.) besides its own weight, it has a considerable tensile strength per unit area. Having tensile strength means that there is a continuous gel structure in the string. The experiment, however, shows that the string twists endlessly in one direction while repeating an asymmetric twisting to the right and left. As it is not twisted by an external force, the lower end of the free-moving protoplasmic string should be moving in the most stable orientation, an orientation that always eliminates the internal distortion and has the lowest potential energy. If being in a gel state means to have a structural continuity, the fact that there is continuous gel structure in the string is not compatible with the fact that the string twists endlessly in one direction. In order to understand such a behaviour in the protoplasm, we have to suppose a shifting of its structural elements (submicroscopic fibrils) against one another. Shifting in the structure may take place by reconversion of junctions among the elements (submicroscopic fibrils) forming them.

How and where such a structural shifting occurs cannot of course be determined merely by this experiment. But it can be concluded from it that, even when protoplasm is in a gel state, the connections among its structural elements cannot be permanent so long as it is living; in other words, these connections are repeatedly undergoing annihilation and reconstruction. As the continuity of the gel must always be preserved if the string is to keep having a definite tensile strength, the shifting in the structure must be occurring locally at different times while preserving the continuity of the gel. When shifting occurs in the gel structure, we might say from the colloidal point of view that, first, the junction forming the gel falls apart locally and solates, then gelates again; that is, the sol-gel transformation occurs in minute spaces. But the concept of simple reconversion of junctions is not sufficient to explain the dominance in rotation in one direction over that in the other. The point which particularly interests us is the implication of the fact that the repetition of a breaking-down and rebuilding process does not occur at random but proceeds according to a special scheme leading to a shift of structural elements in a definite direction. What determines this dynamic anisotropy is a problem to be studied hereafter, but it must be a question that cannot be explained unless we go back to the problem of the structure and orientation of the molecules.

In any case, the occurrence of shifting in the inner structure of the protoplasm is fundamentally the same phenomenon as that of the *streaming* of the protoplasm. According to the prevailing views, the protoplasmic strand is in a gel state, but there occurs solation and shifting in its minute space. Thus it seems that the twisting of the string of the slime mould and the shifting occurring at the sol-gel interface in the rotational streaming are not wholly unrelated phenomena. In general, in physiologically active protoplasm, the connection among its structural elements even at the part which is in a gel state, is not permanent, but rather changes continually in its combination.

6. Causal relation between potential difference and the motive force of the protoplasmic streaming

In their preliminary experiment Kamiya and Abe (1950) determined successively potential differences between two loci of the plasmodium *(Physarum polycephalum)* connected with a single strand of protoplasm suspended in the air. The potential difference thus measured changed according to a rhythmic pattern, the period of which was the same as that of the cyclic back and forth streaming.

As the first step in the experiment in determining the causal relation between these periodic potential variations and the protoplasmic flow, they tried to subject the central part of the connecting strand to local coagulation by dehydration through use of a concentrated sucrose solution. By such a treatment the back and forth flow of the endoplasm along the connecting strand is stopped because of its blockage. Attention must be called in this case to the fact that even when the connecting strand was subjected to such a condition locally, the potential difference between the two masses of protoplasm continued to change as if the streaming were unhindered. They ascertained further that rhythmic changes in potential difference could be seen even when a part of the connecting strand was entirely eliminated and replaced by a wet cotton thread. Such facts must mean that the rhythmic potential variation can proceed quite independently of the protoplasmic streaming and that consequently it cannot be regarded as a so-called *streaming potential* resulting from the streaming as has been discussed elsewhere (Seifriz 1937, 1943).

A further striking proof of this conclusion is to be had in the experiment using the double-chamber. For Kamiya and Abe have made clear that, not only when the flow has been stopped by application of the balance-pressure, but also when pressure has on the contrary been added for a short time in such a manner as to accelerate its speed far more than in its natural condition, or so as to reverse the direction of flow, the potential difference between the two protoplasmic masses is not affected in any discernible manner [9]. This shows that the potential difference thus measured is causally independent of the velocity of the endoplasmic streaming.

On the other hand, the fact reported by KAMIYA and ABE that the phase of the DPG wave precedes always that of the EPG wave[10] indicates that the electric potential cannot be the cause generating the driving force of the protoplasmic streaming (cf. Fig. 29). As a further proof of this, the potential difference (0.01–4.0 volts) established artificially between the two compartments for a short period of time does not exert any noticeable effect upon the wave pattern of DPG and its polarity-line. Therefore any attempt to ascribe the flow in the slime mould to the local potential difference, or to explain the flow as a phenomenon of *electroendosmosis,* must be discarded in the light of these findings.

In his study on correlations between protoplasmic streaming and bioelectric potential of a slime mould, *Physarum polycephalum,* IWAMURA (1949) pointed out that the correlations between the two are not always maintained when several substances, *e. g.,* carbon dioxide, quinine, NaOH, HCl, etc., are applied.

The results of our subsequent experiments (unpublished) also show that, when the plasmodium is brought under low oxygen tension or treated with several metabolic inhibitors, the electric potential and the motive force of the streaming often manifest different modes of reaction. In such cases the parallelism no longer exists between EPG and DPG.

At any rate, the facts obtained so far have clearly indicated that even though a high correlation is seen under normal conditions between the potential difference and the protoplasmic flow in the slime mould, it does not mean that a direct relation of cause and effect exists between them. The nature of the intimate relationship between the two is yet to be probed.

7. The mechanochemical system of the protoplasmic streaming

a) Properties of myosin B in the myxomycete plasmodium

In the physiology of the protoplasmic streaming, the point that has been most discussed for many years and is still most obscure of all, is the question of knowing just what the mechanism is through which the energy ob-

---

[9] Only when the pressure is kept applied to one compartment so that a considerable part of the endoplasm is dislocated from one compartement to the other, the potential difference between the two parts of the plasmodium can be altered in such a way that the part facing the lower pressure, *i. e.,* the part which is expanded by the inflowing endoplasm, becomes lower in electric potential than the part facing higher pressure KISHIMOTO 1957). KISHIMOTO called this interesting phenomenon "pressure effect". It has, however, little to do with the direct effect of pressure but with translocation of the endoplasm brought about by the uneven pressure.

[10] KISHIMOTO (1958 b) suspected that the phase lag of the potential wave behind the motive force wave might be due to the "pressure effect" induced by the application of balance-pressure. His argument, however, is irrelevant, since KISHIMOTO's pressure effect manifests itself only when translocation of the protoplasm occurs, while under the state of dynamoplasmometry the flow is suspended through the application of the balance-pressure and the translocation of protoplasm is just eliminated.

tained by metabolism produces the mechanical phenomenon of streaming. What has particularly been drawing the attention of researchers of late about the mechanochemical system of protoplasmic streaming is the actomyosin-like protein whose presence was known in myxomycete plasmodium.

LOEWY (1952) first discovered in the plasmodium of *Physarum polycephalum* the protein system which responds to ATP just as muscle myosin B (actomyosin) does. The crude extract obtained by him with 1.2 M KCl containing 0.1 M K_2HPO_4 (pH 8.1–8.3) exhibits marked changes in viscosity on addition of ATP. Later the properties of the active principle, which specifically responds to ATP, were studied in detail by Ts'o *et al.* (1956 a, b, 1957 a, b) and NAKAJIMA (1956, 1957). The former authors obtained the active material through successive fractionation by ammonium sulfate and the latter through extraction with WEBER-EDSALL solution (0.6 M KCl, 0.01 M Na_2CO_3, 0.04 M $NaHCO_3$) followed by purification by dilution procedure. All of these authors used the plasmodia of the same myxomycete, *Physarum polycephalum.* Just as does muscle myosin B, the structural protein obtained from myxomycete plasmodia also undergoes superprecipitation in the presence of ATP.

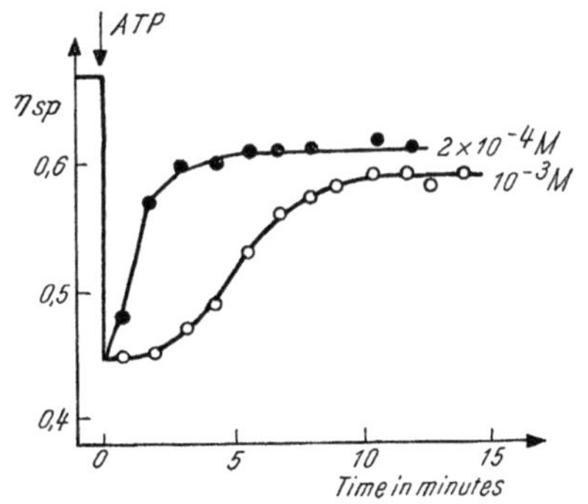

Fig. 81. The effect of ATP on the viscosity of the extract from myxomycete plasmodium with 0.5 M KCl (0.05 M histidin puffer, pH 6.6). 15° C. (NAKAJIMA 1956.)

Fig. 81 shows how the viscosity of the solution changes with time when ATP was added to the extract purified by NAKAJIMA. The ordinate is the specific viscosity of the extract and the abscissa shows the time. Viscosity change under ATP such as presented in the figure shows a time course that is identical in its basic character with that of muscle myosin B. Such a response to ATP being highly characteristic of myosin B, we may infer that a contractile protein, that is similar to myosin B of muscle, may also be present in the myxomycete plasmodium. Ts'o *et al.* referred to this plasmodial myosin B as "myxomyosin."

Ts'o, EGGMAN, VINOGRAD (1957 b) made extensive studies on the physicochemical attributes of this protein by viscosity, sedimentation, birefringence and electron microscope measurements. It was shown that the "myxomyosin" molecule is a long rigid rod with an average length of 4000–5000 Å and a diameter of 70 Å. The molecular weight was calculated to be 6,000,000 on the basis of viscosity and sedimentation data. It is concluded that a ribonucleic acid solution is contained in the active concentrate obtained by salting out (Ts'o *et al.* 1956 b), but it is believed not to be essential to the interaction with ATP (Ts'o *et al.* 1957 b).

Studies conducted by NAKAJIMA (1956, 1957) further made the following facts clear:

1. AMP has no effect upon the viscosity of the plasmodial myosin B.

2. Adding pyrophosphate causes a certain decrease in viscosity, but applying Mg^{++} further increases this change.

3. Contrary to M^{++}, Ca^{++} restores to the control value the viscosity that was decreased by pyrophosphate.

4. Ca^{++} and Mg^{++} have no effect on viscosity when added singly.

5. When Ca^{++} is added simultaneously with ATP, the duration of the lowering of viscosity is shortened; in the case of Mg^{++}, it is lengthened.

6. This purified preparation possesses ATPase activity.

Item 5. is due to the fact that added ATP is destroyed more rapidly when Ca^{++} is present, which stimulates the ATPase activity of myosin B, and the reverse is true in the case of Mg^{++}.

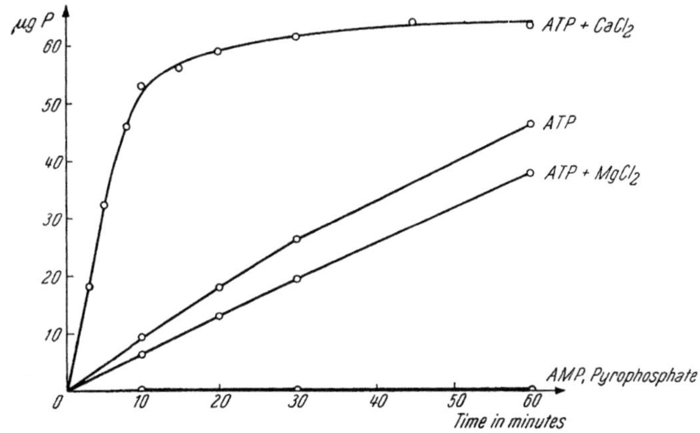

Fig. 82. The dephosphorylation of ATP and the related substances by plasmodial myosin B under different conditions.

K^+: 0.5 M; $CaCl_2$, $MgCl_2$: 3.3×10^{-3} M; ATP: 0.69×10^{-3} M; AMP: 0.83×10^{-3} M; pyrophosphate: 0.85×10^{-3} M. ATPase activity: ATP: 6.6 μg. P/min./mg. N; ATP + $CaCl_2$: 44.7 μg. P/min./mg. N; ATP + $MgCl_2$: 4.5 μg. P/min./mg. N.

(NAKAJIMA 1957.)

Fig. 82 shows how inorganic phosphate is released from ATP by plasmodial myosin B with or without calcium and magnesium. From this we understand that phosphoric acid of AMP and of pyrophosphate are not hydrolysed and further that ATPase activity of myosin B is stimulated to nearly 7 times the control value by Ca^{++} and is inhibited by 30% by Mg^{++}.

All the above behaviours of myosin B obtained from slime moulds correspond well to the properties of myosin B of muscles. This fact together with the behaviour of living plasmodium when exposed to ATP (p. 74), leads to the assumption that myosin B is the substance playing an essential part in mechanochemical coupling in the plasmodium. There is a high possibility that the contraction and relaxation of these protein molecules or the change in their aggregate condition brought about by ATP plays an essential role in the mechanism of the generation of the motive force for the protoplasmic streaming in the plasmodium.

The periodical contraction of this protein induces a partial change in the inner pressure of a slime mould, and this causes the sol endoplasm to be carried passively to the part with lower inner pressure. Such an idea, which has been held by early workers and stressed by Seifriz (1937) when the existence of myosin B as such was not demonstrated, may be a possible explanation of the streaming in slime moulds. More will be discussed about this matter later (p. 170).

b) Properties of the glycerinated cell

Whether it is proper or not to try to understand the various kinds of protoplasmic streaming by reducing them to contraction of protein molecules or change in their aggregation pattern is a matter that cannot be determined till further research is done. In this connection, the recent finding of Takata (1958) is interesting.

In order to investigate whether there are actomyosin-like structural proteins in the *Acetabularia* cell, Takata applied the procedure of glycerination to it and investigated how this glycerinated material responds to ATP. After the cell was kept in 20% glycerol buffered with 1×10^{-2} M Na-K phosphate (pH 7.0) for 15–18 hours at 0^0 C. the material was washed with 0.1 M $KCl + 1 \times 10^{-2}$ M Na-K phosphate buffer. When a solution composed of 1×10^{-3} M $MgCl_2 + 1 \times 10^{-3}$ M ATP is added to the glycerinated cell, a transient motion is observable in a protoplasmic mass in a state of gel. This movement of the dead protoplasm treated with glycerol involves accompanying changes in its aggregate conditions and lasts about 2 minutes, but it does not occur with AMP or pyrophosphate. It is clear that such a movement is not a phenomenon which is comparable with the protoplasmic streaming in a living cell. But it is interesting that no movement is brought about in the glycerinated protoplasmic mass if the KCl concentration is raised above 0.3 M. Furthermore this movement is induced only in the presence of a small amount (*e. g.*, 10^{-3} M) of $MgCl_2$ or $CaCl_2$. These facts bear a striking resemblance to the contraction of glycerinated muscle (Bozler 1954) and suggest the existence of a myosin-like protein, which responds to ATP specifically, in the protoplasm of *Acetabularia*. It is further inferred that it may be the proteins in question which play important parts in the mechanochemical coupling in the protoplasmic streaming.

It is probable that the protein system which is akin to muscle myosin B is of wide occurrence in a cell exhibiting protoplasmic streaming. Granting that contractile proteins are the basis of protoplasmic movement in many cases, the question further remains as to what structural relationships those contractile elements have when protoplasmic motion of a specific type takes place.

X. Theories

The history of theories concerning the mechanism of the protoplasmic streaming dates from the time when the streaming itself was discovered. During more than a century, a variety of energies have been thought of as the causes generating the streaming.

SEIFRIZ (1943) surveyed the past theories dividing them in eleven categories; these try to explain the protoplasmic streaming in terms of the following energies or processes: 1. surface tension, 2. hydration, 3. osmosis, 4. sol-gel reversibility, 5. myelin processes, 6. coacervates, 7. autonomous propulsion of particles, 8. kinetic energy, 9. magnetism, 10. electrical forces, and 11. contractility.

Of these theories, some are to be discarded in our judgment today, but there are some which seem to be adequate in connection with certain particular phenomena and appear to be pointing in the right direction theoretically. None of them, however, seems to be decisive and to have sufficient experimental basis to establish it as something definite.

When we discuss the mechanism of protoplasmic streaming, the first problem to be dealt with is to find out at what part of a cell or of a protoplasmic system the force causing the flow is generated; a further question is the mode of application of that force. In short, both the *dynamic organisation* of the protoplasmic system and the *nature of the energy* must be known in order to elucidate the phenomenon. We shall consider at first the mechanism of a few reprensentative types of protoplasmic streaming separately.

1. Rotational streaming

Contractility, surface tension, hydration-dehydration, electric charge etc. have been postulated as direct causes driving the protoplasm, but the problem is where and how such energies are harnessed so that the ceaseless rotational motion is established. The foregoing pages carry facts which lead us to conclude that the force responsible for protoplasmic rotation resides at the interface between cortical plasmagel and endoplasmic sol. There have been, however, opinions that differ from this view. In the following we shall consider some of the representative theories.

a) Action of endoplasm-vacuole interface

From the standpoint of surface tension theory, the interface between cell sap and endoplasm was once thought to be the site of the driving force (BERTHOLD 1886, cf. also LUNDEGÅRDH 1922). Indeed, the fact that active rotation is usually found in a cell having a well-developed central vacuole inclines us to postulate an important role on the part of the vacuole in the mechanism of the streaming, but this possibility is denied in the light of the fact described in the foregoing that protoplasmic streaming is maintained in a cell the whole space of which is filled with endoplasm (p. 139). Under natural conditions cases are also known in which the total contents of the spherical cells rotate around the center; for example, the rotation occurring in the cysts of *Aethalium* and *Euglena,* in spore mother cells of *Phascum* etc. (cf. HOFMEISTER 1867, p. 46). In a very young 'leaf' cell of *Chara major,* small scattered vacuoles distinctly seen through staining with neutral red are observed to be carried together with the rotating cytoplasm, their deformation, separation, or fusion occurring constantly (CAZALAS 1930). Vacuoles can play no active part in the streaming in this case. The streaming movements in the egg of *Sabellaria* (HARRIS 1935), in the body of *Paramecium* (HOSOI 1937) etc., may also be regarded as being one of the examples of streaming occurring independently of the vacuole. Thus it is clear enough that the existence of vacuoles is by no means a *sine qua non* condition for the protoplasmic rotation.

b) Action of cell surface

WENT (1938) postulated that the whole mechanism of the protoplasmic streaming is seated at the outer protoplasmic layers. The reasons which led him to this rather unusual hypothesis are the facts shown by THIMANN and SWEENEY (1937) and SWEENEY and THIMANN (1938), that there are some discrepancies between the effect of auxin on growth and that on protoplasmic streaming [11]. No matter how auxin affects the protoplasmic streaming in the *Avena* coleoptile, however, it is difficult to accept WENT's hypothesis if only because we do not know how the whole mechanism, seated outside the semipermeable protoplasmic membrane and the immobile ectoplasmic layer, can produce the streaming inside the cell. WENT himself gave no explanation of the physical basis of the mechanism accounting for the streaming. No evidences are known in support of this hypothesis in other materials.

c) Inherent property of streaming protoplasm

In contrast to the above views, not a few authors are inclined to believe that the force responsible for the streaming is located in the streaming protoplasm itself. Authors, who are unanimous in the idea that the streaming is an inherent property of the endoplasm and is not the result of the play of forces at its surface or somewhere else, are not always unanimous in their opinions as to the nature of the force operating in the protoplasm.

According to EWART (1903) the rate of flow in *Chara* is greatest in the central layers of the endoplasm and not at its inner surface. The streaming

[11] If no sugar (*e. g.* fructose) is added, the auxin effect on the streaming is transitory but not on growth; further, at a certain concentration of IAA the streaming is retarded, whereas the growth is promoted. Another fact shown by THIMANN and SCHNEIDER (1938) is that the effects of equimolar indoleacetic or of potassium indoleacetic solutions are the same on growth, whereas the acid is more effective than the salt in producing the streaming. Went thought that the difference in pH between the two solutions as to their effect on the streaming, is due to the difference in their acidity, since the buffering action of the cell is not effective at or outside of the semipermeable protoplasmic membrane. Thus he arrived at the following conclusion: "Protoplasmic streaming is a surface phenomenon, maintained by energy derived from respiration. The whole mechanism is seated at the outer protoplasmic layers, and on account of this location is only poorly provided from within with sugars or other energy-supplying substances, the semi-permeable membrane being located between cell contents and motor mechanism. This explains why buffering of the cell contents does not influence the streaming, while the external pH does so readily. It also explains why the effect of applied auxin is almost immediate, since it does not have to penetrate into the cell."

appears to be absent at its outer surface. Although the work of KAMIYA and KURODA (1956) does not support EWART's statement, it was the above observation, which led EWART to the belief that the energy for the motion is generated in the endoplasm. EWART thought that the motion of the endoplasm, in the last analysis, is a surface phenomenon. If the endoplasm is assumed to be a fluid replete with a multitude of minute particles, and if the lowering of surface tension were to take place on the same side of all the particles in an emulsion, they would move in this direction. In his review article, TIEGS (1928) remarked that EWART's hypothesis seemed to be the only hypothesis of streaming in plant cells that could be entertained at his time. We have, however, no evidence showing the existence in the endoplasm of the spherical bodies postulated in EWART's theory.

It seems to the author that there is also a logical fallacy in EWART's theory, because, under the conditions assumed by him, the particles and their surrounding fluid (hyaloplasm) must move in opposite directions contrary to his belief that the particles drag the surrounding medium.

FREY-WYSSLING (1947/49, 1955) is also of the opinion that the driving force of protoplasmic rotation is located within the streaming protoplasm. In place of surface tension, he ascribed the driving force to the rhytmic contractility of individual protein fibers. Contraction is to be understood not in terms of the folding of polypeptide chains but of the shortening of chains of a larger unit which is brought about through change in the shape of mutually linked macromolecules. By the slackening of a contracted gel, the original form of macromolecules is regained and at the same time lateral cohesion bonds of globule chains arranged in parallel disappear, with the result that elastic gel is transformed again into a structureless plasmasol. All these phenomena, which happen on the submicroscopic level, are certainly not recognizable by observing the protoplasmic streaming; what we see is only particles which are embedded in this system and are carried passively. As FREY-WYSSLING (1955) himself admits, a difficulty involved in this theory is to explain how bodily streaming rather than transmission of contractile waves can be established. In order to help understand a possible sequence of events in the endoplasm giving rise to its bodily rotational streaming, FREY-WYSSLING (1955, p. 199) presented a theoretical diagram.

The idea postulating the contractility of fibrous protein molecules as a physical basis of the rotation or circulation in plant cells is not new. It has been presented by SCHMIDT (1941, 1942/43), by LOEWY (1949), and recently by KISHIMOTO (1958b).

The hypotheses of MARSLAND (1939) and OSTERHOUT (1952) also have something in common with the foregoing theoretical suggestions in that they likewise attribute the seat of the motive force to the streaming protoplasm itself. The difference lies in the nature of the energy accounting for the streaming.

In place of the contractility of proteins, OSTERHOUT (1952) believes that electrostriction is responsible for the production of the streaming. There are, however, few experimental evidences in support of OSTERHOUT's statements.

Further, if we were to follow his logic, what moves would seem to be water in the interstitial space of protoplasm rather than protoplasm *en masse.*

As mentioned already, a high hydrostatic pressure exhibits a marked effect on the sol-gel equilibrium and the streaming of protoplasm (see p. 98). As gelation reactions of protoplasm are accompanied by a small increase in its volume, MARSLAND (1939) thought that the sol-gel reaction may possibly provide protoplasmic streaming with mechanical energy. He suggested a probable sequence of sol-gel reaction in the cytoplasm which might bring forth the rotational streaming, but MARSLAND states in his later paper (1942) that "the question cannot be answered decisively until the focal points where solation and gelation take place have been localized more definitely."

d) Action of sol-gel interface

A way of testing the validity of a theory of rotational streaming is to decide whether it can explain the velocity distribution exhibited by the cell. As already discussed rather fully in Chapter IX, the only logical deduction from the intracellular distribution of velocity studied in *Nitella* (Figs. 5, 6, and especially Fig. 68) seems to the author to be that a sliding force is produced at the sol-gel interface, or the interface between cortical gel and endoplasmic sol. The importance of the sol-gel interface for the protoplasmic streaming has been suggested by several authors who worked mainly on the cells of *Characeae* from different points of view (HÖRMANN 1898, LINSBAUER 1929, BRECKHEIMER-BEYRICH 1954, JAROSCH 1956 b, KAMIYA and KURODA 1956).

In the coenocytic tubes of *Siphonales,* especially in *Acetabularia,* we notice that protoplasm and chloroplasts flow only on definite tracks along the cortical layer (p. 10). The relation between the movement of the protoplasm or chloroplasts and these tracks reminds us of the relation between the creeping motion of fibroblasts in tissue culture and fibrin fibers (cf. GARBER 1953). It is possible that these parallel tracks on the cortical gel in *Acetabularia* serve as a trellis guiding the movement of the endoplasm just as the fibrin fibrils guide the excursions of cytoplasm of the fibroplast. These tracks are not seen as separate striations in *Nitella* or *Chara* cells, yet it is possible that a structure of a similar nature exists on the cortical layer of the cells exhibiting rotation.

As stated before (p. 150) it is a significant discovery of JAROSCH (1956 a, b, 1957) that the fibrils, which can move independently and are ready to form visible bundles, exist in isolated protoplasmic drops obtained from *Chara.* Small particles coming across the action sphere of the fibrils cease their Brownian motion and slide along the fibrils. The motion of the fibrils themselves and that of the particles are always opposite. This fact leads us to conclude that minute counter current is produced in the direct proximity of the moving fibrils. If we suppose that these fibrils, capable of independent motion when free, are attached to the inner surface of the cortical gel with definite orientation, we get an image of the endoplasm in contact with it receiving an impulse to slide through its interaction with the cortical gel; the endoplasm thus moves in a direction parallel to these fibrils over the cortical gel.

Once the sliding force is accepted as existing at the sol-gel interface, the next problem is the mechanism of how a sliding force is established there. Although we have as yet hardly any experimental evidence which permits us to decide on the events occurring there, we can make some theoretical suggestions about possible mechanisms of shift.

Among existing theories on the protoplasmic flow, the concept of a reaction surface suggested by LOEWY (1949) fits nicely into the above view. Fundamentally, LOEWY's idea is similar to that of SCHMIDT or of FREY-WYSSLING in that all these ideas presuppose the rhythmic contraction of linear protein molecules and transience of their lateral bonds. LOEWY assumed that the motile force is located in certain contractile protein constituents of the stream which are elongated and are parallel to the line of flow. These proteins are attached locally and transiently to what he called a "reaction surface" external to the stream by a mobile and weak bond. When the shifting of bonds takes place with proper sequence combined with contraction and relaxation of the molecule, it can creep on the reaction surface in a manner like the movement of a worm; the contractile molecule is thus comparable to the caterpillar body, and the shifting bonds to the caterpillar legs. If we substitute the inner surface of the cortical gel for a reaction surface and postulate the "caterpillar" molecules at the outer edge of the endoplasm, we can image to ourselves a picture of sliding. But this is a conjecture. The nature of the bond in question is of course not known, although LOEWY supposes it to be a hydrogen bond. There is a possibility that cyclic conversion of —S—S— and —SH HS— plays also a role, since SH-reagent para-chloromercuribenzoic acid (PCMB) rapidly causes cessation of protoplasmic streaming in *Nitella* and other cells, cessation which is reversed by cysteine (ABE 1959, see p. 127).

In considering the mechanism of sliding at the sol-gel interface, a recent theory of muscle contraction is very instructive. H. E. HUXLEY showed that there is in the myofibrils a double array of filaments which slide into each other during contraction. According to HANSON and HUXLEY (1955) sliding occurs between actin filaments and myosin filaments. For the mechanism of this sliding, WEBER (1958) suggested a possible series of chemical reactions leading simultaneously to splitting of inorganic phosphate from ATP and to a shifting of the actin filaments past the molecules of myosin A-filaments. In WEBER's scheme SH-group of myosin plays also an important part. The principle of the process, which WEBER presented as a hypothesis, may also be adopted as a hypothetical image of the sliding involved in the protoplasmic streaming.

Another entirely different way of accounting for the mechanism of shift between sol and gel layers is to resort to the concept of interfacial energy. If we assume a cyclic change in surface tension at the outer edge of the endoplasmic layer, and if we further assume that the change is transmitted as waves along the longitudinal axis of the cell—somewhat comparable to the metachronal rhythm on a ciliated tissue—, the endoplasm must be moved along the interface in the same direction as that of the surface tension waves. Though there is, of course, no positive evidence for such a line of thought, it is not impossible that this holds true. In the assumed changes in surface tension, we should like to emphasize the special role played by the phosphatides, which are likely to be rich at sol-gel interface. The conclusion arrived at by KAMIYA and KURODA (1958 b) (see p. 106) that the viscosity at the sol-gel interface is lower than the endoplasmic sol itself, is in favour of this view. If we consider the interface as consisting of phosphatide coacervates, this agrees well with the theory of BOOIJ and BUNGENBERG DE JONG (1956) and the concept of SCARTH (1942). Unfortunately, however, we do not know anything as to how and why the changes in surface tension occur, and what the spatial and chronological controls are that work upon those hypothetical changes to produce the one way streaming. Favourable points supporting the surface tension theory may be found in the fact that the protoplasmic streaming is highly susceptible to the influence of various fat solvents, which are surface-active, and that the absolute value of the motive force for rotational streaming is very small (see p. 40).

2. Streaming in transvacuolar protoplasmic strand

The motion of or in a protoplasmic strand crossing the vacuole, the motion which is seen in the case of circulation (p. 7), is another important object of study in protoplasmic streaming. PFEIFFER (1953) found that the change in the streaming rate of the protoplasm in the transvacuolar strand (*Curcurbita, Tradescantia* and *Allium*) is closely correlated with streaming dichroism. Namely, the streaming dichroism (Δk) increases according as the streaming of the protoplasm forming the strand is accelerated. This fact may well be interpreted to the effect that there are fibrous elements in the strand and the acceleration of the streaming is causally related to their improved orientation. Thus the findings of PFEIFFER support the hypotheses of SCHMIDT, FREY-WYSSLING, LOEWY etc., all of which postulate contractile fibrous molecules of proteins as the prime mover for the streaming.

If we assume that protoplasmic fibrils, such as observed by JAROSCH exist in the protoplasmic strand pulled across the vacuole taking a more or less parallel orientation, the motion of the strand or the particles in it can well be understood. Probably, in the case of circulation, these fibrils are detached partially from the cortical gel and are moving more or less independently in their longitudinal direction. These fibrils seem to be important components of the transvacuolar protoplasmic strand and are responsible for its motion.

3. Streaming in myxomycete plasmodium

Among various types of protoplasmic flow, that in the myxomycete plasmodium is the one in which its physical and chemical nature is so far best known. As already mentioned, theories postulating electric force as the cause of protoplasmic streaming, or the attempt to explain the streaming in terms of electroendosmosis, seem untenable in the light of the facts reported by Kamiya and Abe (1950).

Recently important biochemical facts have been accumulated which enable us to compare the protoplasmic streaming in the plasmodium with muscle contraction. They may be recapitulated as follows:

1. Analytical demonstration of ATP in the plasmodium (Takeuchi and Hatano, see p. 78).

2. Proof for the presence of proteins, with properties akin to those of myosin B (actomyosin) through susceptibility to viscosity change under the effect of ATP (Loewy, Ts'o *et al.*, Nakajima, see p. 161).

3. Augmentation of the motive force caused by application of ATP from the outside or by microinjection (Kamiya *et al.*, Takata etc. see. p. 74).

4. Parallelism of plasmodial myosin B with muscle myosin B in the effect of Ca^{++} and Mg^{++} on the ATPase activity (Nakajima, see p. 162).

Engelhardt (1957) presented several criteria which can help us to decide whether basic mechanisms similar to those of muscular contraction are involved in other forms of motility. The above facts known about myxomycete plasmodia fulfil most of the criteria enumerated by him. Thus we may consider that a far-reaching similarity of the basic enzymatic and mechanochemical systems of the protoplasmic streaming in the myxomycete plasmodium and muscular motion is firmly established. What is unknown still is the physical structure responsible for the rhythmic protoplasmic streaming caused by the interaction of plasmodial myosin B and ATP.

We have already shown that the flow in the well-differentiated strand is caused passively by a difference in internal pressure (see p. 142). There is no evidence in support of the view that the motive force resides in the flowing endoplasm. The mechanism of the protoplasmic streaming in the slime mould is therefore to be sought in terms of how the internal pressure is changed locally.

If we tentatively postulate rhythmic contraction and relaxation of the contractile protein in the cortical gel of the plasmodium, we get a picture reminding us of the relationship between the pulsation of a heart and the passive blood flow. Whether or not such a picture is helpful, however, is still uncertain, since the state of flow in a fan-like plasm expanse, or pseudopod, is much more complicated than a mere sac of plasmagel filled with plasmasol. A vigorous and well-differentiated course of endoplasmic flow in a plasmodial strand is led into the fan-like expanse where it is divided into many minute streams embedded in the plasmagel. At the terminal region, the sol and gel parts are intricately interwoven and demarcation between the two is not clearly observable. Yet one would find it difficult, say Winer and Moore (1941), to believe that a high degree of organization is not present there when seeing the streaming material rushing through the intricate lattice of channels. Cinematographic studies conducted by Stewart and Stewart (1957) at higher magnifications revealed that individual particles there move against or independently of the main stream. The STEWARTS say that the phenomena are not explained by existing theories of protoplasmic movement.

It is noticed through detailed observation that the moment of reversal of the rhythmic flow does not occur at the same time over a pseudopodium or over a plasmodial strand (cf. Fig. 11); a new reverse flow sometimes commences from the distal end toward which protoplasm has been flowing, or sometimes from the basal end from which protoplasm has been coming. This fact indicates that the endoplasm is driven not only by a force from behind (*vis a tergo*) but by suction from the front (*vis a fronte*).

With an image comparing the plasmodial pseudopod to a simple gel sac with a contractile gel wall filled with endoplasm, it is difficult to understand the force from the front, *i. e.*, the existence of the negative pressure inside. Such a phenomenon, however, can be understood, if we assume that a sliding force is produced at the sol-gel interface of the minute channels imbedded in the cortical gel layer, a force that works alternately with a definite period in opposite directions. This reminds us of the movement of *Bacillaria paradoxa* in that the shifting reverses periodically (see p. 61). At any rate, periodic changes in the internal pressure and also the occurrence of the local negative pressure is understandable through the introduction of the idea of sliding. Though of course a conjecture, this idea, if right, may provide a common background for the streaming in the myxomycete plasmodium and for the rotation, such as found in *Nitella* cell, in that they both are based upon sliding along the gel surface. It is interesting indeed in this connection to note that a recent theory of muscle contraction also postulates sliding of the filaments past one another (cf. HANSON and HUXLEY 1955, WEBER 1958, etc.). It is not improbable that the same kind of force is at play both in protoplasmic streaming and muscle contraction.

It has been already stated that there are, among plasmodia, some which show a rather unusual behaviour of streaming. In the plasmodium of *Reticulomyxa filosa,* we see several lines of flow which are directed alternately in the opposite directions (see p. 14). A pressure theory fails to explain such a type of flow. It is mainly this type of streaming which led SEIFRIZ to abandon his former theory (1937, 1942) of contractility. This mode of streaming is, however, again in conformity with the view locating the seat of the motive force at the sol-gel interface.

Now that the presence of myosin-like contractile proteins in the myxomycete plasmodium has become an established fact, the problem yet to be solved is how they are correlated to the dynamic organization producing the motive force, whether through contraction of the cortical gel or sliding at the sol-gel interface, or something else.

4. Independent motion of protoplasmic particles

In what is referred to as protoplasmic streaming in a cell of higher plants especially in agitation and circulation, there are cases in which it is difficult to decide whether the moving particles are carried along with the streaming ground plasm or whether they are moving independently in or on the quiescent ground plasm. In the staminal hair of *Tradescantia*, for instance, we notice that each particle within one and the same strand of protoplasm can move against the streaming of another, or a particle may overtake another particle on one and the same thin strand. No particle changes its direction and speed on the way until it reaches the pole it is aiming at. Bünning (1935, 1953) supposed that the motion represents a sort of electrophoresis of the electrically charged particles.

According to Seifriz (1953), the motion of a particle is established through "depolarization wave." Such waves can be recognized not only in active nerves and muscles, but in protoplasmic strands, and they would thus cause the protoplasmic movement. "If a wave of depolarization travels along the surface, this will have the same effect as if we had pulled electrodes along the axis. The precise mechanism is not known; but it is only necessary to assume that a local depolarization is generated which runs parallel to the fibre axis. If, instead of a muscle fiber, we have a thread of protoplasm within a cell, then a wave of depolarization giving rise to a field parallel to the axis would carry particles and fluids with it. ... Impulses travel back and forth, rhythmically or spasmodically, moving the cytoplasmic fluid and the particles in it." In other words, the hypothesis of Seifriz implies that "the propagation of forces *identical to those which traverse nerve fibers*" (Seifriz 1955) is responsible for the movement of protoplasmic particles. The particles, he says, do not move under their own power nor are they necessarily carried by the protoplasmic stream.

Beautiful examples of the movement of particles which appear to be independent of the stream of the ground plasm, are to be had in protozoa, especially in pseudopodal threads of *Foraminifera* and *Radiolaria* (Schultze 1863, Schneider 1906). Heidenhain (1907, 1911) concluded that the particles are carried by the "smallest waves" of the surrounding plasm and Schmidt (1941) represented Heidenhain's idea in a diagram in terms of the contraction waves of protein fibers.

Andrews (1955) studied some minute movements in the ciliated protozoa *Folliculinidae* and *Stentoridae*. In these organisms food enters from the oesophagus into a food vacuole. The mechanism as to how food can be carried to the food vacuole is a problem which must be closely related with the motion of the protoplasmic particles and which is still left to be solved (cf. Bolzer 1922, 1924).

We have already described how chloroplasts and nuclei rotate or move independently in the isolated endoplasmic drop of *Nitella* or *Chara* (p. 148). They can well be interpreted again as being due to the sliding force produced at the surface of the moving bodies. The reason why they move "independently" is because they have a special gel structure similar to that of the inner surface of the cortical gel of *Nitella* or *Chara* on their surface; attachment of the fibrils such as observed by Jarosch is postulated to exist on their surfaces, fibrils which are responsible for producing the shearing force. If the gel phase is free, their independent motion ensues, whereas if the gel phase is fixed, the endoplasm streams. Therefore both the independent motion and the streaming of the ground plasm may be reduced again to the problem of generation of the counter-shifting force.

All in all, the pivotal point of the mechanism common to apparently different types of protoplasmic motions seems to be focused on the as yet unsolved problem of how the shifting force is established at the boundary of sol and gel phases.

In the field of muscle physiology, TERU HAYASHI and his associates (1952, 1953) constructed an actomyosin fiber which contracts and performs mechanical work upon addition of ATP in the presence of some electrolytes (HAYASHI 1952, 1953; HAYASHI and ROSENBLUTH 1952, 1953).

If we likewise consider making a model of protoplasmic streaming with the natural components of protoplasm in analogy to the above, it will have to consist of two phases, sol and gel. If we can induce such a system to undergo either streaming of the sol phase or independent motion of the gel particles on addition of ATP, it will surely give a powerful clue to the mechanism of protoplasmic streaming and other cellular motions.

XI. Concluding Remarks

In the previous chapters, we have attempted to survey our knowledge on protoplasmic streaming up to the present. The results of observations by a great number of workers with different materials, methods and aims do not necessarily agree nor are their interpretations always in accord. As to the physiological meaning and function of protoplasmic streaming there are also still many obscure points. Nevertheless, the attempt to elucidate the mechanism and nature of this phenomenon has a great significance in helping us to understand the intrinsic character of the living matter itself.

In the present survey, we have been discussing in parallel the streaming within the dermatoplast, which does not involve locomotion or changes in external form, and the streaming in the slime mould, which does involve them. We know that they differ considerably in their mode of streaming and their reaction to metabolic inhibitors. In case of the slime mould, all the experimental results obtained heretofore support the view that the source of the movement lies in the interaction between contractile proteins and ATP. This is an interesting fact inasmuch as it provides us with a basis for supposing that protoplasmic streaming in the slime mould is a phenomenon that stands on a common biochemical basis with muscle contraction and movements of flagella and cilia. Many authors for many years have been inclined to believe that all the diverse forms of biological motility have a similar fundamental mechanism, and recently ENGELHARDT (1958) expressed this principle tersely as "unity among diversity." Data on the slime mould give substantial support to this view.

There is, however, naturally no reason why the fundamental mechanism for the diverse types of motions should necessarily be the same. Among the various types of motions referred to as protoplasmic streaming, it may be that there are some in which the chemical energy does not go through the changes in molecular configuration of proteins or in their aggregation pattern but is transformed into electric, osmotic or interfacial energy, and secondarily causes the movement. For instance, we may regard it as a fact that the tidal streaming seen in the hyphae of *Phycomycetes* is closely connected with osmosis, and there are still problems about the "autonomous" movement of particles in cytoplasm.

When, as in case of *Nitella* and other plant cells, shifting occurs in the interface of sol and gel so that the endoplasm undergoes endless rotational streaming in one direction, its dynamic structure is intriguing. In this case too, it might become possible to explain the mechanism of the streaming in terms of the peculiar and orderly behaviour of protein molecules at the plane of shifting. As a matter of fact the behaviour of the *Acetabularia* cell (pp. 78 and 163) points to the importance of the role of contractile protein and ATP for protoplasmic streaming. But, on the other hand, we also have to keep in mind the possibility of surface-chemical reactions as the cause of the streaming, such as coordinated change in interfacial energy at the sol-gel boundary or change in affinity between endoplasm and cortical gel. For the time being the mechanism of shift at the sol-gel interface must be left in abeyance until further evidences are accumulated.

Leaving for future study the cause of the mechanism producing ceaseless shift between sol and gel phases, we realize that there are in the living cells various yet unelucidated kinematic phenomena which are dealt with separately from protoplasmic streaming but which nevertheless seem to be closely related to the protoplasmic streaming in respect to their physico-chemical background. The behaviour of the kinetochore in nuclear division, the movement of the sperm head reaching the egg nucleus in the egg cytoplasm in fertilization, the "spontaneous" movement of various cytoplasmic particles and organelles, the movement of chloroplasts turning toward light etc. are some of them. If these phenomena are in the last analysis also a matter of shifting, they again have a problem in common with the protoplasmic streaming.

Generally, the shifting of microscopic structures must be constantly occurring in the space within the metabolically active protoplasm. Such a minute structural shifting of protoplasm on the submicroscopical level, which is in many cases not organized and not coordinated on the microscopical level and hence is not recognizable, might have a significant meaning for the metabolic process in the living system.

In order that this minute shifting may be perceived as rotation, the sliding surface must be supported by more or less fixed gel structures. Otherwise there is no foothold for the sliding to develop into a unified streaming. When the gel phase is in the form of granules or in a suspended thread, there may be free movement of these, but no rotational streaming can be brought about. If a protoplasmic system solates completely under certain circumstances, such as under high hydrostatic pressure (p. 98), there no longer occurs either protoplasmic streaming or any other physiological movement. It may seem paradoxical that the viscosity lowers and the streaming stops, but the motive force for the streaming is not generated without gel support.

When we consider that even muscles, which have become highly differentiated as a physiological machine doing mechanical work, originally came from primitive protoplasmic systems, we realize that it is rather natural that muscle-like chemical properties are seen also in protoplasmic systems that present a much lower degree of specialization than muscles—an interesting fact indeed from the standpoint of comparative physiology. When we take a general view of many modes of movement and consider the generation of mechanical force in living organisms from the standpoint of comparative physiology, it seems after all that the predominant feature of protoplasmic streaming is based upon the mechanochemical system represented by myosin-like proteins which have by themselves ATPase activity. It looks as if nature had a predilection for using actomyosin-like protein and ATP as a mechanochemical system for carrying out mechanical work in many cells besides muscles. At the present state of our knowledge, which is indeed still very rudimentary, it may be tentatively assumed, without denying other possibilities in special cases, that this is the basis of the mechanism of movement widely used in protoplasm in general. It is through it that are formed different dynamic structures which determine the various types of protoplasmic movement and modes of streaming.

Acknowledgements

The author owes not a small part of this survey to the researches that have been conducted together with, or by, his co-workers and the frequent discussions held with them have been most helpful in writing these pages. Especially Dr. Kiyoko Kuroda spared no efforts to help the author in many ways, taking original photographs for this work, locating the references, etc. Not a few text-figures were prepared through the help of Drs. Hiromichi Nakajima and Shigemi Abe. Overcoming the linguistic difficulties the writer had to cope with was made possible only through the devoted help of Dr. Miyeko Kamiya. Many of the necessary papers that were not to be found at Osaka University have been made available from elsewhere through the good offices of Mr. Shigeharu Kosai and Miss Sayoko Matsunami of our library.

The author expresses his cordial thanks to all the above people without whose cooperation this work could never have been accomplished.

Bibliography

Abe, S., 1959: Rotational protoplasmic streaming and SH. (Japanese.) Kagaku (Science) **29**, 361—362.

— H. Nakajima, and N. Kamiya, 1957: Simultaneous measurement of respiration and the motive force of protoplasmic streaming in the myxomycete plasmodium. IV. The effect of monoiodoacetic acid. Proc. Jap. Acad. **34**, 697—699.

Adrian, E. D., and B. H. C. Matthews, 1934: The Berger rhythm: potential changes from the occipital lobes in man. Brain **57**, 355—385.

Agardh, C. A., 1826: Über die Anatomie und den Kreislauf der Charen. Nova Acta Acad. Leop.-Carol. Nat. Curios. **13**, 113—162.

Allen, P. J., and W. H. Price, 1950: The relation between respiration and protoplasmic flow in the slime mold, *Physarum polycephalum*. Amer. J. Bot. **37**, 393—402.

Amlong, H. U., 1936: Zur Frage der Wuchsstoffwirkung auf das Wurzelwachstum. Jb. wiss. Bot. **83**, 773—780.

Andersen, N., and B. M. Pollock, 1952: A comparison between the cytoplasmic components in the myxomycete, *Physarum polycephalum*, and in the amoeba, *Chaos chaos*. C. r. Labor. Carlsberg, Sér. Chim. **28**, 247—264.

Anderson, J. D., 1951: Galvanotaxis of slime mold. J. Gen. Physiol. **35**, 1—16.

—, J. V. Butkiewicz, and E. C. Roter, 1953: Ca, Na and K concentrations in the cathodal and anodal portions of *Physarum polycephalum* during Galvanotaxis. Meet. Gen. Physiol. Woods Hole.

Andrews, E. A., 1955: Some minute movements in protoplasm. Biol. Bull. **108**, 121—124.

Andrews, F. M., 1912: Protoplasmic streaming in *Mucor*. Bull. Torrey Bot. Club **39**, 455—499.

Arnon, D. I., 1955: The chloroplast as a complete photosynthetic unit. Science **122**, 9—16.

Arthur, J. C., 1897: The movement of protoplasm in coenocytic hyphae. Ann. Bot. **11**, 491—507.

Audus, L. J., 1953: Plant growth substances. London.

Auger, D., 1931: Relation entre le courant d'action et la cyclose protoplasmique chez *Nitella*. C. r. soc. biol. Paris **108**, 1131—1132.

— 1933: Contribution à l'étude de la propagation de la variation électrique chez les Characées. C. r. soc. biol. Paris **113**, 1437—1440.

— 1939: L'activité protoplasmique des cellules végétales. Actualités Scientifiques et Industrielles **823**. Paris.

Balbach, H., 1936: Über Plasmadifferenzierungen in Plasmodien der Schleimpilze. Protoplasma **26**, 161—180.

Barg, T., 1943: Über die Indifferenzstreifen der Characeen. Protoplasma **37**, 293—299.

Bärlund, H., 1938: Einfluß des Äthyläthers auf die Permeabilität der *Chara*-Zellen. Protoplasma **30**, 70—78.

Beams, H. W., 1949: Some effects of centrifuging upon protoplasmic streaming in *Elodea*. Biol. Bull. **96**, 246—256.

Becker, Z. E., 1936: A comparison between the action of carbonic acid and other acids upon the living cell. Protoplasma **25**, 161—175.

Beikirch, H., 1925: Die Abhängigkeit der Protoplasma-Strömung von Licht und Temperatur und ihre Bedingtheit durch andere Faktoren. Bot. Arch. **12**, 389—445.

Bělehrádek, J., 1925: La viscosité du protoplasma dans ses rapports avec l'activité et le viellissement cellulaire. C. r. soc. biol. Paris **92**, 1070—1072.

— 1929: Sur la signification des coëfficients de température. Protoplasma **7**, 232—255.

— 1935: Temperature and living matter. Protoplasma-Monographien **8**. Berlin.

Berger, J., P. Smith, and G. S. Avery, Jr., 1946: The influence of auxin on respiration of the *Avena* coleoptile. Amer. J. Bot. **33**, 601—604.

Berthold, G., 1886: Studien über Protoplasmamechanik. Leipzig.

Bierberg, W., 1909: Die Bedeutung der Protoplasmarotation für den Stofftransport. Flora **99**, 52—80.

Bierling, R., 1955: Zur Bedeutung der lokalen Mitochondriendichte für die Formkonstanz von Zellen. Protoplasma **44**, 299—306.

Bishop, C. J., V. D. McLaughlin, and D. F. Tapley, 1949: X-ray tolerance of living cells as measured by cytoplasmic streaming. Canad. J. Res., Sect. C **27**, 262—268.

Bonner, J., 1949: Respiration and growth in the *Avena* coleoptile. Amer. J. Bot. **36**, 429—436.

— and R. S. Bandurski, 1952: Studies of the physiology, pharmacology, and biochemistry of the auxins. Ann. Rev. Plant Physiol. **3**, 59—86.

Booij, H. L., and H. G. Bungenberg de Jong, 1956: Biocolloids and their interactions. Protoplasmatologia **1**, 2. Wien.

Bottelier, H. P., 1933: Über den Einfluß des Lichtes auf die Protoplasmaströmung von *Avena*. Proc. Kon. Akad. Wet. Amsterdam **36**, 790—794.

— 1934: Über den Einfluß äußerer Faktoren auf die Protoplasmaströmung in der *Avena*-Koleoptile. Rec. trav. bot. Néerl. **31**, 474—582.

Bottelier, H. P., 1935: Oxygen as limiting factor of the protoplasmic streaming in *Avena* coleoptiles of different ages. Rec. trav. bot. Néerl. **32**, 287—292.
Bozler, E., 1922: Über die Theorie der Protoplasmaströmungen. S. B. Ges. Morph. u. Physiol. München **34**, 58—67.
— 1924: Über die physikalische Erklärung der Schlundfadenströmungen, ein Beitrag zur Theorie der Protoplasmaströmungen. Z. vergl. Physiol. **2**, 82—90.
— 1954: Binding of calcium and magnesium by the contractile elements. J. Gen. Physiol. **38**, 735—742.
Breckheimer-Beyrich, H., 1949: Über die Wirkung zentrifugaler Kräfte auf das Protoplasma von *Nitella flexilis*. Ber. dtsch. bot. Ges. **62**, 55—60.
— 1954: Weitere Erkenntnisse über die Wirkung zentrifugaler Kräfte auf das Protoplasma von *Nitella flexilis*. Ber. dtsch. bot. Ges. **67**, 86—92.
Brown, D. E. S., 1934: The pressure coefficient of "viscosity" in the eggs of *Arbacia punctulata*. J. cellul. a. comp. Physiol. **5**, 335—346.
Brücke, E., 1862: Das Verhalten der sogenannten Plasmaströme in den Brennhaaren von *Urtica urens* gegen die Schläge des Magnetelektromotors. S. B. Wien. Akad. Wiss., math.-naturw. Kl. II, **46**, 35—38.
Buller, A. H. R., 1933: Researches on fungi. 5th ed. London, New York and Toronto.
Bünning, E., 1935: Einige Beobachtungen über die Körnchenströmung in den Staubfadenhaaren von *Tradescantia*. Nebst Bemerkungen über die Verwendbarkeit des Mikropolychromars bei zellphysiologischen Studien. Z. wiss. Mikrosk. **52**, 166—172.
— 1953: Entwicklungs- und Bewegungsphysiologie der Pflanze. Berlin.
Bushee, G. L., 1908: The occurrence and rate of protoplasmic streaming in greenhouse plants. Bot. Gaz. **46**, 50—53.
Buy, H. G. du, and R. A. Olson, 1938 a: Protoplasmic streaming and dynamics of transport through living cells. Biodynamica **36**, 1—18.
— — 1938 b: Protoplasmic streaming, electric potentials and growth in coleoptiles of *Triticum* and *Avena*. Science **87**, 490—491.
— — 1940: The relation between respiration, protoplasmic streaming and auxin transport in the *Avena* coleoptile, using a polarographic microrespirometer. Amer. J. Bot. **27**, 401—413.
Camp, W. G., 1936: A method of cultivating myxomycete plasmodia. Bull. Torrey Bot. Club **63**, 205—210.
— 1937: The structure and activities of myxomycete plasmodia. Bull. Torrey Bot. Club **64**, 307—335.
Cazalas, 1930: Sur l'evolution du vacuome des *Chara* dans ses relations avec les mouvements du cytoplasme. C. r. acad. sci., Paris **190**, 314—317.
Clark, W. G., 1938: Electrical polarity and auxin transport. Plant Physiol. **13**, 529—552.
Cohen, A. L., 1939: Nutrition of the myxomycetes. I. Pure culture and two-membered culture of myxomycete plasmodia. Bot. Gaz. **101**, 243—275.
— 1941: Nutrition of the myxomycetes. II. Relations between plasmodia, bacteria, and substrate in two-membered culture. Bot. Gaz. **103**, 205—224.
Cole, K. S., and H. J. Curtis, 1938: Electric impedance of *Nitella* during activity. J. Gen. Physiol. **22**, 37—64.
Colla, S., 1929: Untersuchungen über Plasma und Plasmaströmung bei Characeen. II. Die Wirkung verschiedener Salze auf die Plasmaströmung. Protoplasma **6**, 438—448.
Commoner, B., and K. V. Thimann, 1941: On the relation between growth and respiration in the *Avena* coleoptile. J. Gen. Physiol. **24**, 279—296.
Cook, S. F., 1929: The effect of sudden changes of temperature on protoplasmic streaming. J. Gen. Physiol. **12**, 793—803.
Corti, B., 1774: Osservazioni microscopiche sulla tremella e sulla circolazione del fluido in una pianta acquajuola. Lucca.
Currier, H. B., 1949: Responses of plant cells to herbicides. Plant Physiol. **24**, 601—609.
Curtis, O. F., 1935: The translocation of solutes in plant. New York.
Dalleux, G., 1940: Recherches sur les plasmodes de deux myxomycètes. Rev. Cytol. Cytophysiol. Végét. **4**, 123—182.
Dehnecke, C., 1881: Einige Beobachtungen über den Einfluß der Präparationsmethode auf die Bewegungen des Protoplasma der Pflanzenzellen. Flora, **39**, 8—14, 24—30.
Denny, F. E., 1926: Hastening the sprouting of dormant potato tubers. Contrib. Boyce Thompson Inst. **1**, 59—66.

Diannelidis, T., und K. Umrath, 1953: Über das elektrische Potential und über den Erregungsvorgang bei dem Myxomyceten *Physarum polycephalum*. Protoplasma **42**, 312—323.
Doi, Y., 1950: Protoplasmic streaming in root hairs of crop plants. (Japanese with English résumé.) Nat. Agr. Exper. Station (Tokyo) Bull. No. **69**, 1—47.
Donné, A., 1838: Note sur la circulation du *Chara*. Ann. Sci. Nat. II Bot. **10**, 346—348.
Dostál, R., 1929: Untersuchungen über Protoplasmamobilisation bei *Caulerpa prolifera*. Jb. wiss. Bot. **71**, 596—667.
Dutrochet, 1837: Observation sur le *Chara flexilis;* modification dans la circulation de cette plante sous l'influence d'un changement de température, d'une irritation mécanique, de l action des sels, des acides et des alcalis, de celle des narcotiques et de l'alcool. C. r. acad. sci., Paris **5**, 775—784.
— 1838: Observations sur la circulation des fluides chez la *Chara fragilis*. Ann. Sci. Nat. II Bot. **9**, 5—38, 65—79.
— et Ad. Brongniart, 1938: Rapport fait à l'Académie des Sciences dans sa séance du 30 avril, par M. Dutrochet, sur un Mémoire du M. Donné, relatif à certains phénomènes de mouvement observés chez le *Chara hispida*. Ann. Sci. Nat. II Bot. **10**, 349—355.
Ebbecke, U., 1936: Über plasmatische Kontraktionen von roten Blutkörperchen, Paramäcien und Algenzellen unter der Einwirkung hoher Drucke. Pflügers Arch. B **238**, 452—466.
Engel, H., 1954: Effects of the injection of snake venom containing lecithinase A into *Physarum polycephalum*. J. cellul. a. comp. Physiol. **44**, 203—210.
Engelhardt, W. A., 1957: Enzymology and mechanochemistry of tissues and cells. Proc. Intern. Symp. Enzyme Chem. (Tokyo) 34—39.
Ewart, J., 1903: On the physics and physiology of protoplasmic streaming in plants. Oxford.
Eymers, J. G., and H. P. Bottelier, 1937: Protoplasmic movement in the *Avena* coleoptile as related to oxygen pressure and age. Proc. Kon. Akad. Wet. Amsterdam **40**, 589—595.
Fetzmann, E., 1957: Rotierende Eigenbewegung der Zellkerne von *Chara foetida*. Anzeig. math.-naturw. Kl. Öst. Akad. Wiss. Jahrg. Nr. **14**, 287—289.
— 1958: Über rotierende Eigenbewegung der Zellkerne und Plastiden bei *Chara foetida*. Protoplasma **49**, 549—556.
Fiedler, H., 1936: Entwicklung und reizphysiologische Untersuchungen an Kulturen isolierter Wurzelspitzen. Z. Bot. **30**, 385—436.
Fitting, H., 1925: Untersuchungen über die Auslösung von Protoplasmaströmung. Jb. wiss. Bot. **64**, 281—388.
— 1927: Untersuchungen über Chemodinese bei *Vallisneria*. Jb. wiss. Bot. **67**, 427—596.
— 1929: Über die Auslösung von Plasmaströmung durch optisch-aktive Aminosäuren. Jb. wiss. Bot. **70**, 1—25.
— 1930: Untersuchungen über die Natur der chemodinetischen Reizung und über die Unterschiedsschwellen für l-Asparagin. Z. Bot. **23**, 328—360.
— 1932: Untersuchungen über die Empfindlichkeit und das Unterscheidungsvermögen der *Vallisneria*-Protoplasten für verschiedene α-Aminosäuren. Jb. wiss. Bot. **77**, 1—103.
— 1933 a: Untersuchungen über den Plasmaströmung auslösenden Reizstoff in den Blattextrakten von *Vallisneria*. Jb. wiss. Bot. **78**, 319—398.
— 1933 b: Untersuchungen über den die Protoplasmaströmung auslösenden Reizstoff der Pflanzen. Naturwiss. **21**, 489—495.
— 1936 a: Über die Auslösung von Protoplasmaströmung bei *Vallisneria* durch einige Histidinverbindungen. Jb. wiss. Bot. **82**, 613—624.
— 1936 b: Über die Auslösung von Protoplasmaströmung durch chemische Wirkstoffe. Forsch. Fortsch. **12**, 160—161.
— 1937: Beiträge zur Physiologie der Protoplasmaströmung in den Blättern von *Vallisneria spiralis*. Ber. dtsch. bot. Ges. **55**, 255—261.
Franck, F., and D. Auger, 1932: Analyse cinématographique des mouvements du protoplasme en rapport avec la variation électrique dans l'excitation chez *Nitella*. C. r. acad. sci., Paris **195**, 1321—1322.
Frank, B., 1872: Über die Veränderung der Lage der Chlorophyllkörner und des Protoplasmas in der Zelle, und deren innere und äußere Ursachen. Jb. wiss. Bot. **8**, 216—303.
Freundlich, H., 1937: Some recent work on gel. J. Physic. Chem. **41**, 901—910.

Frey-Wyssling, A., 1940: The submicroscopic structure of the cytoplasm. J. Roy. microsc. Soc. **60**, 128—139.
— 1947/49: Das Plasmagel. Proc. 6th Intern. Congr. Exper. Cytol. (Exper. Cell Res. Suppl. **1**, 33—42.)
— 1953: Submicroscopic morphology of protoplasm. 2nd English ed. (1st English ed. 1948). Amsterdam and New York.
— 1955: Die submikroskopische Struktur des Cytoplasmas. Protoplasmatologia **2**, A 2. Wien.
Fujii, T., S. Utida, and T. Mizuno, 1955: Reaction of starfish spermatozoa to histidine and certain other substances considered in relation to zinc. Nature **176**, 1068—1069.
— — — and S. Nanao, 1955: Effects of amino acids and some chelating substances on the motility and the oxygen uptake of starfish spermatozoa. J. Fac. Sci. Univ. Tokyo **7**, 335—345.
Garber, B., 1953: Quantitative studies on the dependence of cell morphology and motility upon the fine structure of the medium in tissue culture. Exper. Cell Res. **5**, 132—146.
Geitler, L., 1942: Eine neue Chytridiale, *Zygorhizidium verrcosum* n. sp., und ihre Wirkung auf die Wirtszellen. Arch. Protistenk. **96**, 109—118.
Gibbs, R. D., 1926: The action of ultra-violet light on *Spirogyra*. Proc. a. Trans. Roy. Soc. Canada, 3rd Ser. **20**, Sec. V, 419—426.
Gimesi, N. I., and B. J. Pozsár, 1955: About the physiology of protoplasmic movements. Acta Biol. acad. sci. Hungaricae **6**, 113—132.
Goldacre, R. J., 1952: The folding and unfolding of protein molecules as a basis of osmotic work. Internat. Rev. Cytol. **1**, 135—164.
— and J. J. Lorch, 1950: Folding and unfolding of protein molecules in relation to cytoplasmic streaming, amoeboid movement and osmotic work. Nature **166**, 497—500.
Green, P. B., 1954: The spiral growth pattern of the cell wall in *Nitella axillaris*. Amer. J. Bot. **41**, 403—409.
— and G. B. Chapman, 1955: On the development and structure of the cell wall in *Nitella*. Amer. J. Bot. **42**, 685—693.
Grehn, J., 1932: Untersuchungen über Gestalt und Funktion der Sporangienträger bei den Mucorineen. II. Der Wasser- und Stofftransport. Jb. wiss. Bot. **76**, 167—207.
Haas, J. N., 1953: Effects of some enzymes injected into the slime mold *Physarum polycephalum*. J. cellul. a. comp. Physiol. **41**, 180—200.
Hanson, T., and H. E. Huxley, 1955: The structural basis of contraction in striated muscle. Symposia Soc. Exper. Biol. **9**, 228—264.
Hanstein, J. von, 1880: Einige Züge aus der Biologie des Protoplasmas. Bot. Abh. Gebiet Morph. Physiol. **4**, Heft 2, 1—56, Bonn.
Harris, J. E., 1935: Studies on living protoplasm I. Streaming movements in the protoplasm of the egg of *Sabellaria alveolata* (L.) J. exper. Biol. **12**, 65—79.
Harvey, N., 1942 a: Stimulation of cells by intense flashes of ultraviolet light. J. Gen. Physiol. **25**, 431—444.
— 1942 b: Hydrostatic pressure and temperature in relation to stimulation and cyclosis in *Nitella flexilis*. J. Gen. Physiol. **25**, 855—863.
Hauptfleisch, P., 1892: Untersuchungen über die Strömung des Protoplasmas in behäuteten Zellen. Jb. wiss. Bot. **24**, 173—234.
Hayashi, Teru, 1952: Contractile properties of compressed monolayers of actomyosin. J. Gen. Physiol. **36**, 139—152.
— 1953: Surface-spread protein as a basis for cell structure and cell movement. Amer. Naturalist **87**, 209—227.
— and R. Rosenbluth, 1952: Contraction-elongation cycle of loaded surface-spread actomyosin fibers. J. cellul. a. comp. Physiol. **40**, 495—506.
— — 1953: Mechanochemical properties of fibers of surface-spread actomyosin. Proc. nat. acad. sci. **39**, 1285—1290.
Hayashi, Toshio, 1952: Some aspects of behavior of the protoplasmic streaming in plant cells. Bot. Mag. (Tokyo) **65**, 51—55.
— 1957: Some dynamic properties of the protoplasmic streaming in *Chara*. Bot. Mag. (Tokyo) **70**, 168—174.
— and E. Kamitsubo, 1959: Plasmolysis in *Characea*. Bot. Mag. (Tokyo) **72**, 275—281.
Heidenhain. M., 1907, 1911: Plasma und Zelle. I. 1, I. 2. Jena.

Heilbrunn, L. V., 1928: The colloid chemistry of protoplasm. Protoplasma-Monographien **1**. Berlin.
— 1940: Protoplasm and colloids. The cell and protoplasm (ed. by Moulton) (Lancaster, Pa.), 188—198.
— 1952: An outline of general physiology. 3rd ed. (1st ed. 1937, 2nd ed. 1943). Philadelphia and London.
— and K. Daugherty, 1939: The electric charge of protoplasmic colloids. Physiol. Zool. **12**, 1—12.
Hertel, E., 1904: Über Beeinflussung des Organismus durch Licht, speziell durch die chemisch wirksamen Strahlen. Z. allg. Physiol. **4**, 1—43.
Heymann, E., 1935: Studies on sol ⇌ gel transformation. I. Inverse sol ⇌ gel transformation of methyl-cellulose in water. Trans. Faraday Soc. **31**, 846—864.
— 1936: Studies on sol ⇌ gel transformation. II. Delatometric investigations on iron hydroxide, gelatin, methylcellulose, silicic acid and viscose. Trans. Faraday Soc. **32**, 462—473.
Hill, S. E., 1941: The relation between protoplasmic streaming and the action potential in *Nitella* and *Chara*. Biol. Bull. **81**, 296.
Hilton, A. E., 1908: On the cause of reversing currents in the plasmodia of Mycetozoa. J. Quekett Microsc. Club **10**, 263—270.
Hoffmann-Berling, H., 1954a: Adenosintriphosphat als Betriebsstoff von Zellbewegungen. Biochim. Biophys. Acta **14**, 182—194.
— 1954b: Die Bedeutung des Adenosintriphosphat für die Zell- und Kernteilungsbewegungen in der Anaphase. Biochim. Biophys. Acta **15**, 226—236.
— 1954c: Die Glycerin-Wasser-extrahierte Telophasezelle als Modell der Zytokinese. Biochim. Biophys. Acta **15**, 332—339.
— 1955: Geißelmodelle und Adenosintriphosphat (ATP). Biochim. Biophys. Acta **16**, 146—154.
Höfler, K., 1918: Eine plasmometrisch-volumetrische Methode zur Bestimmung des osmotischen Wertes von Pflanzenzellen. Denkschr. Akad. Wiss. Wien, math.-naturw. Kl. **25**, 99—170.
— 1928: Über Kappenplasmolyse. Ber. dtsch. bot. Ges. **46**, (73)—(82).
— 1934: Kappenplasmolyse und Salzpermeabilität. Z. wiss. Mikrosk. **51** (Küster-Festschr.), 70—87.
— 1939: Kappenplasmolyse und Ionenantagonismus. Protoplasma **33**, 545—578.
— 1940a: Aus der Protoplasmatik der Diatomeen. Ber. dtsch. bot. Ges. **58**, 97—120.
— 1940b: Salzquellung des Protoplasmas und Ionenantagonismus. Ber. dtsch. bot. Ges. **58**, 292—305.
— A. Ziegler und M. Luhan, 1956: Fluorochromierungsstudien mit Uranin. Protoplasma **46**, 322—366.
Hofmeister, W., 1867: Die Lehre von der Pflanzenzelle. Handb. physiol. Bot. **1**, Leipzig.
Holter, H., and B. M. Pollock, 1952: Distribution of some enzymes in the cytoplasm of the myxomycete, *Physarum polycephalum*. C. r. Trav. Labor. Carlsberg, Sér. chim. **28**, 221—245.
Hörmann, G., 1898: Studien über die Protoplasmaströmung bei den Characeen. Jena.
Hosoi, T., 1937: Protoplasmic streaming in isolated pieces of *Paramecium*. J. Fac. Sci. Imp. Univ. Tokyo, Sect. **IV/4**, 299—305.
Howard, F. L., 1931: The life history of *Physarum polycephalum*. Amer. J. Bot. **18**, 116—133.
Hürthle, K., 1915: Eine Methode zur Registrierung der Geschwindigkeit des Blutstroms in den kapillaren Gefäßen. Pflügers Arch. **162**, 422—432.
Ito, S., and F. Yamamoto, 1957: Studies on the physiological responses of some crops in relation to the organic phosphorated and the chlorinated insecticides. I. Influence on the germination of seeds and the seedings of some crops. (Japanese.) J. Agric. Sci. Tokyo Nogyo Daigaku, **3**, 281—302.
Iwamura, T., 1949: Correlations between protoplasmic streaming and bioelectric potential of a slime mold, *Physarum polycephalum*. Bot. Mag. (Tokyo) **62**, 126—131.
— 1952: Electric impedance of the plasmodium of a slime mold, *Physarum polycephalum*. Cytologia **17**, 322—328.
Iwanami, Y., 1952: Physiological researches of pollen. I. On the protoplasmic streaming in the pollen tube. (Japanese.) Bot. Mag. (Tokyo) **65**, 137—144.
— 1953: Physiological researches on pollen. II. On the protoplasmic streaming in the pollen. (Japanese.) Bot. Mag. (Tokyo) **66**, 49—54.

IWANAMI, Y., 1956: Protoplasmic movement in pollen grains and tubes. Phytomorphology 6, 288—295.
JAHN, A., 1934: Über Wachstum, Plasmaströmung und vegetative Fusionen bei *Humaria leucoloma* Hedw. Z. Bot. 27, 193—250.
JAHN, E., 1932: Die Organe des Plasmodiums. Myxomycetenstudien Nr. 14. Ber. dtsch. bot. Ges. **50 a**, 367—399.
JANSE, J. M., 1890: Die Bewegungen des Protoplasma von *Caulerpa prolifera*. *Jb.* wiss. Bot. **21**, 163—284.
JAROSCH, R., 1956 a: Aktiv bewegungsfähige Plasmaelemente und Chloroplastenrotation bei Characeen. Anzeig. math.-naturw. Kl. Österr. Akad. Wiss. Jahrg. Nr. **6**, 58—60.
— 1956 b: Plasmaströmung und Chloroplastenrotation bei Characeen. Phyton (Argentina) **6**, 87—107.
— 1956 c: Die Impulsrichtungsänderungen bei der Induktion der Protoplasmaströmung. Protoplasma **47**, 478—486.
— 1957: Zur Mechanik der Protoplasmafibrillenbewegung. Biochim. Biophys. Acta **25**, 204—205.
JOHNSON, G. T., and C. MOOS, 1956: Succinoxidase activity of mitochondria from myxomycete plasmodia. J. cellul. a. comp. Physiol. **48**, 343—252.
JOSING, E., 1901: Der Einfluß der Außenbedingungen auf die Abhängigkeit der Protoplasmaströmung vom Licht. Jb. wiss. Bot. **36**, 197—228.
JOST, L., 1929: Einige physikalische Eigenschaften des Protoplasmas von *Valonia* und *Chara*. Protoplasma **7**, 1—22.
JURIŠIĆ, P. J., 1922: Osmotischer Druck und Protoplasmaströmung in Pflanzenzellen. Z. allg. Physiol. **20**, 100—110.
— 1925: Die Geschwindigkeitsänderung der Protoplasmaströmung in den Pflanzenzellen. Acta bot. Inst. bot. Univ. Zagreb **1**, 25—31.
KAISERLEHNER, E., 1939: Über Kappenplasmolyse und Entmischungsvorgänge im Kappenplasma. Zugleich ein Beitrag zur Kenntnis der Salznekrose des Cytoplasmas. Protoplasma **33**, 579—601.
KAMIYA, N., 1937: Untersuchungen über die Wirkung des elektrischen Stromes auf lebende Zellen. I. Das Verhalten der mitotischen Figur unter der Wirkung des Gleichstromes. Cytologia Fujii-Jub. Vol., 1036—1042.
— 1939: Die Rhythmik des metabolischen Formwechsels der Euglenen. Ber. dtsch. bot. Ges. **57**, 231—240.
— 1940: Control of protoplasmic streaming. Science **92**, 462—463.
— 1942: Physical aspects of protoplasmic streaming. The structure of protoplasm (ed. by SEIFRIZ. Monogr. Amer. Soc. Plant Physiol. (Ames, Iowa), 199—244.
— 1943: Motive force of the protoplasmic streaming. I, II, III. (Japanese.) Kagaku (Science) **13**, 351—354, 390—395, 438—440.
— 1950 a: The rate of the protoplasmic flow in the myxomycete plasmodium. I. Cytologia **15**, 183—193.
— 1950 b: The rate of the protoplasmic flow in the myxomycete plasmodium. II. Cytologia **15**, 194—204.
— 1950 c: The protoplasmic flow in the myxomycete plasmodium as revealed by a volumetric analysis. Protoplasma **39**, 344—357.
— 1953: The motive force responsible for protoplasmic streaming in the myxomycete plasmodium. Ann. Rep. Sci. Works Osaka Univ. **1**, 53—83.
— 1955: The chemistry of protoplasmic motion. (Japanese.) The chemistry of life phenomena (ed. by ASHIDA, EGAMI and KIKKAWA) (Tokyo), 22—60.
— 1957: The physicochemistry of the protoplasmic streaming. (Japanese.) Sym. Biophys. Chem. **2**, 25—39.
— and S. ABE, 1950: Bioelectric phenomena in the myxomycete plasmodium and their relation to protoplasmic flow. Colloid Sci. **5**, 149—163.
— — and H. NAKAJIMA, 1957 a: Simultaneous measurement of respiration and the motive force of protoplasmic streaming in the myxomycete plasmodium. I. Outline of the method. Proc. Jap. Acad. **33**, 206—210.
— — — 1957 b: Simultaneous measurement of respiration and the motive force of protoplasmic streaming in myxomycete plasmodium. II. The effect of cyanide. Proc. Jap. Acad. **33**, 407—419.
— — — 1958: Simultaneous measurement of respiration and the motive force of protoplasmic streaming in the myxomycete plasmodium. III. The effect of oxygen tension. Proc. Jap. Acad. **34**, 530—533.
— and K. KURODA, 1956 a: Artificial modification of the osmotic pressure of the plant cell. Protoplasma **46**, 423—436.

KAMIYA, N., and K. KURODA, 1956 b: Velocity distribution of the protoplasmic streaming in *Nitella* cells. Bot. Mag. (Tokyo) **69**, 544—554.
— — 1957 a: Cell operation in *Nitella*. I. Cell amputation and effusion of the endoplasm. Proc. Jap. Acad. **33**, 149—152.
— — 1957 b: Cell operation in *Nitella*. II. Behaviour of isolated endoplasm. Proc. Jap. Acad. **33**, 201— 205.
— — 1957 c: Cell operation in *Nitella*. III. Specific gravity of the cell sap and endoplasm. Proc. Jap. Acad. **33**, 403—406.
— — 1958 a: Studies on the velocity distribution of the protoplasmic streaming in the myxomycete plasmodium. Protoplasma **49**, 1—4.
— — 1958 b: Measurement of the motive force of the protoplasmic rotation in *Nitella*. Protoplasma **50**, 144—148.
— — 1958 c: Cell operation in *Nitella*. IV. Tension at the surface of the effused endoplasmic drops. Proc. Jap. Acad. **34**, 435—438.
— and H. NAKAJIMA, 1955: Some aspects of rhythmicity of the protoplasmic streaming in the myxomycete plasmodium. Jap. J. Bot. **15**, 49—55.
— — and S. ABE, 1957: Physiology of the motive force of protoplasmic streaming. Protoplasma **48**, 94—112.
— and W. SEIFRIZ, 1954: Torsion in a protoplasmic thread. Exper. Cell Res. **6**, 1—16.
— and M. TAZAWA, 1956: Studies on water permeability of a single plant cell by means of transcellular osmosis. Protoplasma **46**, 394—422.
KELLER, I. A., 1890: Über Protoplasma-Strömung im Pflanzenreich. Diss. Zürich.
KELSO, J. M., and J. S. TURNER, 1955: Protoplasmic streaming in *Tradescantia*. I. The effects of indoleacetic acid and other growth-promoting substances on streaming. Austral. J. biol. Sci. **8**, 19—35.
KIENITZ-GERLOFF, F., 1893: Protoplasmaströmungen und Stoffwanderung in der Pflanze. Im Anschluß an HAUPTFLEISCH's „Untersuchungen über die Strömung des Protoplasmas in behäuteten Zellen". Bot. Zeitung **51**, 36—42.
KINOSHITA, S., 1938: Über die Wirkung des Wuchsstoffes auf die Bewegung des Plasmodiums. Bot. Mag. (Tokyo) **52**, 492—497. (Japanese with German résumé.)
— 1939: Über den Einfluß des Wuchsstoffs auf die Protoplasmaströmung bei den Myxomyceten-Plasmodien. Bot. Mag. (Tokyo) **53**, 175—180. (Japanese with German résumé.)
KIRCHHEIMER, F., 1933: Protoplasma und Wundheilung bei *Phycomyces*. Planta **19**, 574—606.
KISHIMOTO, U., 1958 a: Rhythmicity in the protoplasmic streaming of a slime mold, *Physarum polycephalum*. I. A statistical analysis of the electric potential rhythm. J. Gen. Physiol. **41**, 1205—1222.
— 1958 b: Rhythmicity in the protoplasmic streaming of a slime mold, *Physarum polycephalum*. II. Theoretical treatment of the electric potential rhythm. J. Gen. Physiol. **41**, 1223—1244.
— and H. AKABORI, 1957: On the relation between protoplasmic streaming and action potential in *Nitella*. 22nd Ann. Meet. Bot. Soc. Japan.
KITCHING, J. A., and M. H. PIRENNE, 1940: The influence of low tensions of oxygen on the protoplasmic streaming of Myxomycetes. J. cellul. a. comp. Physiol. **16**, 131—133.
KLEMM, P., 1895: Desorganisationserscheinungen der Zelle. Jb. wiss. Bot. **28**, 627—700.
KOK, A. C. A., 1931: Über den Einfluß der Plasmarotation auf den Stofftransport. Proc. Kon. Akad. Wet. Amsterdam **34**, 918—929.
— 1932: Über den Transport von Kaffein und $LiNO_3$ durch parenchymatisches Gewebe. Proc. Kon. Akad. Wet. Amsterdam **35**, 3—12.
— 1932/33: Über den Transport körperfremder Stoffe durch parenchymatisches Gewebe. Rec. trav. bot. Néerl. **30**, 23—139.
KOPETZKY-RECHTPERG, O., 1938: Mitteilungen über eine Protoplasmastruktur der Desmidiale *Netrium digitus* und über einige Zellinhaltskörper dieser Alge. Protoplasma **31**, 607—611.
— 1949: Zellbau und Zelleinschlüsse bei Conjugaten, besonders Desmidiales. Protoplasma **39**, 106—112.
KOSHTOYANTZ, H. S., 1957: Fundamentals of comparative physiology. **2**. (Russian.) Moscow.
KRETZSCHMAR, P. N., 1904: Über Entstehung und Ausbreitung der Protoplasmaströmung infolge von Wundreiz. Jb. wiss. Bot. **39**, 273—304.
KRISZAT, G., 1954: Die Wirkung von Para-Chloromercuribenzoesäure und ATP auf Amöben (*Chaos chaos*). Ark. Zool. **6**, 195—201.

KUMAGAYA, S., 1950: Cytoplasmic streaming in pollen grains. (Japanese.) Bot. Mag. (Tokyo) **63**, 52.
KURODA, K., 1956: Recherches sur la relation entre la pression osmotique de la cellule et le courant protoplasmique. Cytologia **21**, 200—207.
— 1958: Sur le courant protoplasmique dans le fragment de protoplasme d'un Myxomycète. C. r. s. biol., Paris **152**, 392—394.
KÜSTER, E., 1910: Über Inhaltsverlagerungen in plasmolysierten Zellen. Flora **100**, 267—287.
— 1933 a: Protoplasmabewegungen in zentrifugierten Zellen. Ber. Oberhess. Ges. Natur-Heilk. Gießen **15**, 194—218.
— 1933 b: Hundert Jahre *Tradescantia*. Jena.
— 1933 c: Dellen und Löcher im Protoplasma lebender Pflanzenzellen. Protoplasma **19**, 443—451.
— 1939: Beobachtungen an Plasmaexplantaten von *Bryopsis*. Z. wiss. Mikrosk. **56**, 113—147.
— 1956: Die Pflanzenzelle. 3. Aufl. (1. Aufl. 1935, 2. Aufl. 1951). Jena.
KUSUNOKI, S., 1940: Neue Beobachtungen über Protoplasmaströmung bei *Hydrilla verticillata*. Cytologia **10**, 539—544.
— 1943 a: Beiträge zur Kenntnis der Protoplasmaströmung. Über verschiedene Arten der Rotation und einige andere Erscheinungen bei *Hydrilla verticillata*. (Japanese with German résumé.) Bot. Mag. (Tokyo) **57**, 237—245.
— 1943 b: Über Protoplasmaströmung in den bei Dauerkulturen von *Nitella* sp. gebildeten Rhizoiden. (Japanese with German résumé.) Bot. Mag. (Tokyo) **57**, 274—276.
LAKON, G., 1914: Beiträge zur Kenntnis der Protoplasmaströmung. Ber. dtsch. bot. Ges. **32**, 421—429.
LAMBERS, M. H. R., 1925: The influenc of temperature on protoplasmic streaming of *Characeae*. Proc. Kon. Akad. Wet. Amsterdam **28**, 340—346.
— 1926: Temperatuur en protoplasmastrooming. Diss. Utrecht.
LAUTERBACH, L., 1921: Untersuchungen über die Beeinflussung der Protoplasmaströmung der Characeen durch mechanische und osmotische Eingriffe. Beih. bot. Cbl. **38**, I. Abt., 1—52.
LEGLER, F., und H. SCHINDLER, 1939: Zentrifugierungsversuche an Diatomeenzellen. Protoplasma **33**, 469—473.
LEPOW, S. S., 1938: Some reactions of slime mould protoplasm to certain alkaloids and snake venoms. Protoplasma **31**, 161—179.
LETTRÉ, H., 1951: Zellstoffwechsel und Zellteilung. Naturwiss. **38**, 490—496.
— 1952: Weitere Beobachtungen über die Abhängigkeit der Zellform und -bewegung von der Adenosintriphosphorsäure. Naturwiss. **39**, 266.
— 1954: Weitere Beobachtungen über die Abhängigkeit von Zellbewegungen von der Adenosintriphosphorsäure. Naturwiss. **41**, 306.
— M. ALBRECHT und R. LETTRÉ, 1951 a: Verhalten von Fibroblasten unter aeroben und anaeroben Bedingungen. Naturwiss. **38**, 504—505.
— — — 1951 b: Zur Auslösung von Plasmabewegungen in Ruhezellen. Naturwiss. **38**, 505.
— und A. SCHLEICH, 1955: Zur Bedeutung der Adenosintriphosphorsäure für Formkonstanz und Formänderungen von Zellen. Protoplasma **44**, 314—321.
LEWIS, W. H., 1942: The relation of the viscosity changes of protoplasm to ameboid locomotion and cell division. The structure of protoplasm (ed. by SEIFRIZ). Monogr. Amer. Soc. Plant Physiol. (Ames, Iowa), 163—197.
LINDENBEIN, 1937: Beobachtungen über Bewegung und Struktur des Protoplasmas in Pflanzenzellen. Cytologia Fujii-Jub. Vol., 311—320.
LINSBAUER, K., 1929: Untersuchungen über Plasma und Plasmaströmung an *Chara*-Zellen. I. Beobachtungen an mechanisch und operativ beeinflußten Zellen. Protoplasma **5**, 563—621.
— 1933: Untersuchungen über Plasma und Plasmaströmung an *Chara*-Zellen. V. Untersuchungen des Protoplasmas mittels der Ausflußmethode. Protoplasma **18**, 554—595.
LOCQUIN, M., 1949: L'utilisation du rayonnement β du radium pour l'étude des courants protoplasmiques dans les plasmodes de myxomycètes. Trav. Bot. (Alger) **2**, 209—214.
LOEWY, A. G., 1949: A theory of protoplasmic streaming. Proc. Amer. Philosoph. Soc. **93**, 326—329.
— 1950: Protoplasmic streaming under anaerobic conditions in a myxomycete. J. cellul. a. comp. Physiol. **35**, 151—153.

LOEWY, A. G., 1952: An actomyosin-like substance from the plasmodium of a myxomycete. J. cellul. a. comp. Physiol. **40**, 127—156.
LOPRIORE, G., 1895: Über die Einwirkung der Kohlensäure auf das Protoplasma der lebenden Pflanzenzelle. Jb. wiss. Bot. **38**, 531—626.
— 1897: Azione dei raggi X sul protoplasma della cellula vegetale vivente. Nuova Rassegna. Catania.
LOU, C. H., S. H. WU, W. C. CHANG, and L. M. SHAO, 1957: Intercellular movement of protoplasm as a means of translocation of organic material in garlic. Scientia Sinica **6**, 139—157.
LUERSSEN, C., 1868: Über den Einfluß des roten und blauen Lichtes auf die Plasmaströmung. Diss. Jena.
LUNDEGÅRDH, H., 1922: Zelle und Cytoplasma. Linsbauers Handb. Pflanzenanat. **1**. Berlin.
MAGENOT, G., 1934: Recherches cytologiques sur les plasmodes de quelques Myxomycètes. Rev. Cytol. et Cytophysiol. veg. **1**, 19—68.
MARCY, B., 1937: Effect of ethylene chlorhydrin and thiourea on *Elodea* and *Nitella*. Plant Physiol. **12**, 207—212.
MARSLAND, D. A., 1936: The cleavage of *Arbacia* eggs under hydrostatic compression. Anatom. Rec. **67**, 38.
— 1938: The effects of high hydrostatic pressure upon cell division in *Arbacia* eggs. J. cellul. comp. Physiol. **12**, 57—70.
— 1939: The mechanism of protoplasmic streaming. The effect of high hydrostatic pressure upon cyclosis in *Elodea canadensis*. J. cellul. a. comp. Physiol. **13**, 23—30.
— 1942: Protoplasmic streaming in relation to gel structure in the cytoplasm. The structure of protoplasm (ed. by SEIFRIZ). Monogr. Amer. Soc. Plant Physiol. (Ames, Iowa), 127—161.
— and D. E. S. BROWN, 1942: The effect of pressure on sol-gel equilibria with special reference to myosin and other protoplasmic gels. J. cellul. a. comp. Physiol. **20**, 295—305.
MARTIN, G. W., 1940: The myxomycetes. Bot. Rev. **6**, 356—388.
MEYER, A., 1920: Die Plasmabewegung, verursacht durch eine geordnete Wärmebewegung von Molekülen. Ber. dtsch. bot. Ges. **38**, 36—43.
MOORE, S. M., 1888 a: Studies in vegetable biology. — III. The influence of light upon protoplasmic movement. Part I. J. Linn. Soc. Bot. **24**, 200—251.
— 1888 b: Studies in vegetable biology. — IV. The influence of light upon protoplasmic movement. Part II. J. Linn. Soc. Bot. **24**, 351—389.
MOTOKAWA, H., 1943: Über den Mechanismus der Entstehung und Hemmung des Hirnpotentials. Tohoku J. Exper. Med. **45**, 297—308.
MÜLLER, O., 1896 a: Die Ortsbewegung der Bacillariaceen. III. Ber. dtsch. bot. Ges. **14**, 54—64.
— 1896 b: Die Ortsbewegung der Bacillariaceen. IV. Ber. dtsch. bot. Ges. **14**, 111—128.
MÜNCH, E., 1930: Die Stoffbewegungen in der Pflanze. Jena.
NADSON, G. A., et E. J. ROCHLIN, 1934: L'effet des rayons X sur le protoplasme, le noyau et le chondriome de la cellule végétale d'après les observations sur le vivant. Protoplasma **20**, 31—41.
NAGAO, M., and Y. OHWAKI, 1954: Relations of the auxin-induced growth of coleoptiles to respiration and fermentation in the *Oryza* seedling growing under water. Sci. Rep. Tohoku Univ., 4th Ser. Biol. **20**, 189—194.
NÄGELI, C., 1860: Beiträge zur wissenschaftlichen Botanik. II. Heft. Leipzig.
— und C. CRAMER, 1855: Pflanzenphysiologische Untersuchungen. I. Heft. Zürich.
— und S. SCHWENDENER, 1877: Das Mikroskop. Theorie und Anwendung desselben. 2. Aufl. Leipzig.
NAKAJIMA, H., 1956: Some properties of a contractile protein in the slime mold. (Japanese.) Seitano-Kagaku (Medical Science) **7**, 256—259.
— 1957: A contractile protein in the myxomycete plasmodium. 22nd Ann. Meet. Bot. Soc. Japan.
NAUSS, R. N., 1949: *Reticulomyxa filosa* gen. et sp. nov., a new primitive plasmodium. Bull. Torrey Bot. Club **76**, 161—173.
NICHOLS, S. P., 1922: Methods of healing in some algal cells. Amer. J. Bot. **9**, 18—27.
— 1925: The effect of wound upon the rotation of the protoplasm in the internodes of *Nitella*. Bull. Torrey Bot. Club **52**, 351—363.
— 1930: The effect of chloroform upon the rotation in the internodes of *Nitella*. Bull. Torrey Bot. Club **57**, 153—162.

NOETHLING, W., und E. ROCHLIN, 1931: Über Photodinese im kurzwelligen Ultraviolett. Planta **14**, 112—131.

NORTHEN, H. T., 1946: Effects of various agents on the structural viscosity of *Elodea* protoplasm. Plant Physiol. **21**, 148—154.

NOTHMANN-ZUCKERKANDL, H., 1915: Über die Erregung der Protoplasmaströmung durch verschiedene Strahlenarten. Ber. dtsch. bot. Ges. **33**, 301—313.

OHTA, J., 1952 a: Experimental studies on the protoplasmic streaming in the myxomycete plasmodium I. Some observations on the motive force of protoplasmic streaming. Cytologia **17**, 210—218.

— 1952 b: Experimental studies on the protoplasmic streaming in the myxomycete plasmodium II. The effects of some respiratory poisons and reduced oxygen tensions on the motive force of protoplasmic streaming. Cytologia **17**, 300—310.

— 1954: Studies on the metabolism of the myxomycete plasmodium. J. Biochem. **41**, 489—497.

— 1958: Experimental studies on the protoplasmic streaming in the myxomycete plasmodium III. The effects of amino acids and some chelating substances on the motive force of protoplasmic streaming. Cytologia **23**, 232—238.

OLSON, R. A., and H. G. DUBUY, 1940: Factors influencing the protoplasmic streaming in the oat coleoptile. Amer. J. Bot. **27**, 392—401.

OSTERHOUT, W. J. V., 1952: Some aspects of protoplasmic motion. J. Gen. Physiol. **35**, 519—527.

— and S. E. HILL, 1931: Electrical variations due to mechanical transmission of stimuli. J. Gen. Physiol. **14**, 473—485.

— — 1933: Anesthesia produced by distilled water. J. Gen. Physiol. **17**, 87—98.

PEASE, D. C., 1940: Hydrostatic pressure effects upon protoplasmic streaming in plasmodium. J. cellul. a. comp. Physiol. **16**, 361—375.

PEKAREK, J., und R. FÜRTH, 1931: Über die Richtung der Protoplasmaströmung in benachbarten *Elodea*-Blatt-Zellen. Protoplasma **13**, 666—700.

PÉTERFI, T., und G. YAMAHA, 1931: Die Wirkung des mechanischen Drucks auf das Protoplasma der *Nitella*-Zelle. Protoplasma **12**, 279—286.

PFEFFER, W., 1877: Osmotische Untersuchungen. Leipzig.

— 1904: Pflanzenphysiologie. 2. Aufl. Leipzig.

PFEIFFER, H., 1931: Kleine Beiträge zur Bestimmung des IEP von Protoplasten. IV. Strömungsgeschwindigkeit und Zentrifugalverlagerung des rotierenden Plasmas in *Nitella*-Zellen nach Vorbehandlung mit Puffergemischen. Protoplasma **14**, 90—96.

— 1953: Veränderungen des Strömungsdichroismus an Protoplasmasträngen pflanzlicher Dauerzellen. Ber. dtsch. bot. Ges. **66**, 2—5.

POP, L. J. J., 1938: Protoplasmic streaming in relation to spiral growth of *Phycomyces*. Proc. Roy. Acad. Amsterdam. **41**, 661—672.

PRICE, W., and P. ALLEN, 1942: Studies on respiration and flow in the myxomycete, *Physarum polycephalum*. Amer. J. Bot. **29**, 15—16.

PRINGSHEIM, N., 1881: Über Lichtwirkung und Chlorophyllfunktion in der Pflanze. Jb. wiss. Bot. **12**, 288—437.

REINKE, J., 1871: Über den Einfluß farbigen Lichtes auf lebende Pflanzenzellen. Bot. Zg. **29**, 790—793, 797—802.

RIETSEMA, J., 1950: Action and penetration of growth substances with special reference to *Avena* coleoptile sections. Diss. Utrecht.

ROBBINS, W. J., 1926: The isoelectric point for plant tissue and its importance in absorption and toxicity. Univ. Missouri Studies **1**, 3—60.

ROMIJN, G., 1931: Über den Einfluß der Temperatur auf die Protoplasmaströmung bei *Nitella flexilis*. Proc. Kon. Acad. Wet. Amsterdam **34**, 289—296.

ROSS, E., 1934: Action of respiratory catalysts and inhibitors on oxygen consumption by *Nitella*. Proc. Soc. Exp. Biol. Med. **32**, 64—66.

— 1938: The effects of sodium cyanide and methylene blue on oxygen consumption by *Nitella clavata*. Amer. J. Bot. **25**, 458—463.

SACHS, J., 1882: Vorlesungen über Pflanzenphysiologie. Leipzig.

SAKAMURA, T., und H. KANAMORI, 1935: Über die Wirkung der Essigsäure, des Ammoniaks und ihrer Salze auf das Protoplasma des Wurzelhaares. J. Fac. Sci. Hokkaido Imp. Univ., Ser. V **4**, 65—98.

SANDAN, T., 1955: Physiological studies on growth and morphogenesis of the isolated plant cell cultured in vitro. I. General feature on the growth and morphogenesis of the internodial cell of *Characeae*. Bot. Mag. (Tokyo) **68**, 274—280.

SANDAN, T., and T. OGURA, 1957: Physiological studies on growth and morphogenesis of the isolated plant cell cultured *in vitro*. III. The effects of pH, auxin and metabolic inhibitors. Bot. Mag. (Tokyo) **70**, 125—130.

SCARTH, G. W., 1927: The structural organization of plant protoplasm in the light of micrurgy. Protoplasma **2**, 189—205.

— 1942: Structural differentiation of cytoplasm. The structure of protoplasm (ed. by SEIFRIZ). Monogr. Amer. Soc. Plant Physiol. (Ames, Iowa), 99—107.

SCHMIDT, W. J., 1941: Die Doppelbrechung des Protoplasmas und ihre Bedeutung für die Erforschung seines submikropischen Baues. Ergebn. Physiol. **44**, 27—95.

— 1942/43: Über das Wesen der Kontraktilitätserscheinungen. Mikrokosmos **36**, 41—43.

SCHNEIDER, K. C., 1906: Plasmastruktur und -bewegung bei Protozoen und Pflanzenzellen. Arb. Zool. Inst. Wien **16**, 99—118.

SCHORR, L., 1935: Anthozyan im Protoplasma. Beiträge zur zellenphysiologischen Methodik. Z. wiss. Mikrosk. **52**, 369—379.

— 1939: Über das Schicksal von Plasmaexplantaten von *Phycomyces*. Arch. exper. Zellforsch. **23**, 338—350.

SCHRANK, A. R., 1947: Electrical and curvature responses of the *Avena* coleoptile to transversely applied direct current. Bioelectric fields and growth (ed. by LUND) (Austin), 217—231.

— 1948: Electrical and curvature responses of the *Avena* coleoptile to transversely applied direct current. Plant Physiol. **23**, 188—200.

— 1949: Influence of longitudinally applied direct current on the electrical polarity and curvature of the *Avena* coleoptile. J. cellul. a. comp. Physiol. **33**, 1—16.

SCHRÖTER, A., 1905: Über Protoplasmaströmung bei Mucorineen. Flora **95**, 1—30.

SCHULTZE, M., 1863: Das Protoplasma der Rhizopoden und der Pflanzenzellen. Leipzig.

SCHULZE, J., 1910: Über die Einwirkung der Lichtstrahlen von 280 $\mu\mu$ Wellenlänge auf Pflanzenzellen. Beih. bot. Cbl. **25**, 30—80.

SCHUMACHER, W., 1936: Untersuchungen über die Wanderung des Fluoresceins in den Haaren von *Cucurbita pepo*. Jb. wiss. Bot. **82**, 507—533.

SCHWEICKERDT, H., 1928: Untersuchungen über Photodinese bei *Vallisneria spiralis*. Jb. wiss. Bot. **68**, 79—134.

SEIFRIZ, W., 1922: A method for inducing protoplasmic streaming. New Phytol. **21**, 107—112.

— 1936: Reaction of protoplasm to radium radiation. Protoplasma **25**, 196—200.

— 1937: A theory of protoplasmic streaming. Science **86**, 397—398.

— 1938: Recent contributions to the theory of protoplasmic structure. Science **88**, 21—25.

— 1939 a: The symmetrical distribution of protoplasmic particles in an electric field. Protoplasma **32**, 536—537.

— 1939 b: Pathological changes in protoplasm. Protoplasma **32**, 538—550.

— 1941: A theory of anesthesia based on protoplasmic behavior. Anesthesiology **2**, 300—309.

— 1942: Some physical properties of protoplasm and their bearing on structure. The structure of protoplasm (ed. by SEIFRIZ). Monogr. Amer. Soc. Plant Physiol. (Ames, Iowa), 245—264.

— 1943: Protoplasmic streaming. Bot. Rev. **9**, 49—123.

— 1952: The rheological properties of protoplasm. Deformation and flow in biological systems (ed. by FREY-WYSSLING) (Amsterdam, New York), 3—156.

— 1953: Mechanism of protoplasmic movement. Nature **171**, 1136—1138.

— 1955: Microscopic and submicroscopic structure of cytoplasm. Handb. Pflanzenphysiol. (ed. by RUHLAND) (Berlin) **1**, 301—339.

— and N. EPSTEIN, 1941: Shock anesthesia in Myxomycetes. Biodynamica **3**, 191—197.

— and M. URAGUCHI, 1941: The toxic effects of heavy metals on protoplasm. Amer. J. Bot. **28**, 191—197.

SENN, G., 1908: Die Gestalts- und Lageveränderungen der Pflanzen-Chromatophoren. Leipzig.

SHINKE, N., 1939: Experimental studies of cell-nuclei. Memoirs Coll. Sci. Kyoto Imp. Univ. Ser. B **15**, 1—126.

SHOWACRE, J. L., and H. G. DUBUY, 1947: The relation of water availability and auxin in the growth of *Avena* coleoptiles and its meaning for a theory of tropisma. Amer. J. Bot. **34**, 175—182.

SIBAOKA, T., and K. ODA, 1956: Shock stoppage of the protoplasmic streaming in relation to the action potential in *Chara*. Sci. Rep. Tôhoku Univ. 4th Ser. Biol. 22, 157—166.
SMIRNOVA, N. A., 1955: Reversible changes of protoplasmic flow in *Elodea* under blocking and recovery of sulfhydril groups. (Russian.) Fiziologia Rastenii **2**, 6, 578—580.
SOBELS, J. C., 1950: Nutrition de quelques myxomycètes en cultures pures et associées et leurs propriétés antibiotiques. Diss. Gouda.
SSAWOSTIN, P. W., 1930: Magnetphysiologische Untersuchungen. I. Die Rotationsbewegung des Plasmas in einem konstanten magnetischen Kraftfeld. Planta **11**, 683—726.
STÅLBERG, N., 1927: Studien über den Zellinhalt von *Nitella opaca*. Bot. Not. 305—322.
STÅLFELT, M. G., 1945: Über die lichtbedingten Hemmungsvorgänge in der Kohlensäureassimilation. Sv. Bot. Tidskr. **39**, 365—395.
— 1946: The influence of light upon the viscosity of protoplasm. Ark. Bot. Stockholm **33**, A, No. **4**, 1—17.
— 1947/49: Effect of heteroauxin and colchicine on protoplasmic viscosity. Proc. 6th Intern. Congr. Exper. Cytol. (Exper. Cell Res. Suppl. **1**, 63—78).
STEWART, P. A., and B. T. STEWART, 1957: Biophysical aspects of protoplasmic streaming (motion picture). Feder. Proc. **16**, 125.
STRUGGER, S., 1926: Untersuchungen über den Einfluß der Wasserstoffionen auf das Protoplasma der Wurzelhaare von *Hordeum vulgare* L. I. S. B. Österr. Akad. Wiss. math.-naturw. Kl. I, **135**, 143—169.
— 1928: Untersuchungen über den Einfluß der Wasserstoffionen auf das Protoplasma der Wurzelhaare von *Hordeum vulgare* L. II. S. B. Österr. Akad. Wiss. math.-naturw. Kl. I, **137**, 453—478.
— 1949: Praktikum der Zell- und Gewebephysiologie der Pflanzen. Berlin.
SWEENEY, B. M., 1941: Conditions affecting the acceleration of protoplasmic streaming by auxin. Amer. J. Bot. **28**, 700—702.
— 1942: The effect of colamine on growth and protoplasmic streaming in *Avena*. Amer. J. Bot. **29**, 793—797.
— 1944: The effect of auxin on protoplasmic streaming in root hairs of *Avena*. Amer. J. Bot. **31**, 78—80.
— and K. V. THIMANN, 1938: The effect of auxins on protoplasmic streaming. II. J. Gen. Physiol. **21**, 439—461.
— — 1942: The effect of auxins on protoplasmic streaming. III. J. Gen. Physiol. **25**, 841—854.
TAKATA, M., 1957: Physiological studies on the protoplasmic streaming in plasmodia by means of microdissection. 22nd Ann. Meet. Bot. Soc. Japan.
— 1958: Protoplasmic streaming in *Acetabularia calyculus*. (Japanese.) Kagaku Science **28**, 142.
TAKEUCHI, I., and S. HATANO, 1955: Energy-rich phosphate compounds in the slime mould and their levels in relation to some external conditions. 20th Ann. Meet. Bot. Soc. Japan.
— — 1956: Acid soluble energy-rich phosphate compounds in the slime mould. II. 21th Ann. Meet. Bot. Soc. Japan.
TASAKI, I., and N. KAMIYA, 1950: Electrical response of a slime mold to mechanical and electrical stimuli. Protoplasma **39**, 333—343.
TAUC, L., 1953: Quelques observations de bioélectricité cellulaire, en particulier chez un myxomycète (*Physarum polycephalum*). J. Physiol. **45**, 232—233.
— 1954: Phénomènes bioélectriques observés dans le plasmode d'un myxomycète (*Physarum polycephalum*). J. Physiol. **46**, 659—669.
TERNETZ, C., 1900: Protoplasmabewegung und Fruchtkörperbildung bei *Ascophanus carneus*. Pers. Jb. wiss. Bot. **35**, 273—312.
THIMANN, K. V., and R. H. LANE, 1938: After-effects of the treatment of seeds with auxin. Amer. J. Bot. **25**, 535—543.
— and C. L. SCHNEIDER, 1938 a: The role of salts, hydrogen-ion concentration and agar in the response of the *Avena* coleoptile to auxins. Amer. J. Bot. **25**, 270—280.
— and B. M. SWEENEY, 1937 a: Action of auxin on protoplasmic streaming. Nature **140**, 807—808.
— — 1937 b: The effect of auxins upon protoplasmic streaming. Gen. Physiol. **21**, 123—135.
TIEGS, O. W., 1928: Surface tension and theory of protoplasmic movement. Protoplasma **4**, 88—139.

Tobias, J. M., and S. Solomon, 1950: Electrically induced polar changes in viscosity in the hyaline protoplasm of *Elodea* with observations on streaming and plastid charge. J. cellul. a. comp. Physiol. **35**, 1—9.
Tomiyama, K., 1954: Cytological studies of resistance of potato plants to *Phytophthora infestans*. 1. The processes of alteration in the host cell produced by invasion of parasite. (Japanese with English résumé.) Res. Bull. Hokkaido Nat. Agri. Exp. Stat. **67**, 28—38.
Treviranus, L. Ch., 1806: Vom inwendigen Bau der Gewächse und von der Saftbewegung in denselben. Göttingen.
Ts'o, P. O. P., J. Bonner, L. Eggman, and J. Vinograd, 1956 a: Observations on an ATP-sensitive protein system from the plasmodia of myxomycete. J. Gen. Physiol. **39**, 325—347.
— L. Eggman and J. Vinograd, 1956 b: The isolation of myxomyosin, an ATP-sensitive protein from the plasmodium of a myxomycete. J. Gen. Physiol. **39**, 801—812.
— — — 1957 a: The interaction of myxomyosin with ATP. Arch. Biochem. Biophys. **66**, 64—70.
— — — 1957 b: Physical and chemical studies of myxomyosin, an ATP-sensitive protein in cytoplasm. Biochem. Biophys. Acta **25**, 532—542.
Umrath, K., 1933: Der Erregungsvorgang bei *Nitella mucronata*. Protoplasma **17**, 258—300.
— 1934: Der Einfluß der Temperatur auf das elektrische Potential, den Aktionsstrom und die Protoplasmaströmung bei *Nitella mucronata*. Protoplasma **21**, 329—334.
— 1953: Über Aktionsstrom und Stillstand der Protoplasmaströmung bei *Nitella opaca*. Protoplasma **42**, 77—82.
Uraguchi, M., 1941: Rhythmic banding in protoplasm. Cytologia **11**, 332—337.
Valkanov, A., 1934: Über die kinetische Energie einiger Zellbestandteile. Protoplasma **20**, 20—30.
Velten, W., 1872: Über die Verbreitung der Protoplasmabewegungen im Pflanzenreiche. Bot. Z. **30**, 645—653.
— 1873: Bewegung und Bau des Protoplasmas. Flora **56**, 81—89, 97—102, 113—128.
— 1876 a: Die Einwirkung der Temperatur auf die Protoplasmabewegung. Flora **59**, 177—182, 193—199, 209—217.
— 1876 b: Aktiv oder passiv? Österr. bot. Z. **26**, 77—88.
— 1876 c: Einwirkung strömender Elektrizität auf die Bewegung des Protoplasma, auf den lebendigen und todten Zellinhalt, sowie auf materielle Theilchen überhaupt. S. B. kais. Akad. Wiss. Wien, math.-naturw. Kl. I. **73**, 343—376.
Vesque-Püttlingen, F. von, 1876: Notiz über Periodicität der Protoplasmaströmung. Bot. Ztg. **34**, 572—575.
Virgin, H. I., 1947/49: Changes in the viscosity of *Spirogyra* cytoplasm under the influence of light. Proc. 6th Intern. Congr. Exper. Cytol. (Exper. Cell Res. Suppl. **1**, 79—84.)
— 1948: Changes in the viscosity of the cytoplasm of *Helodea densa* Casp. during continuous illumination. Physiol. Plant. **1**, 147—155.
— 1949: The relation between the viscosity of the protoplasm, the plasma flow and the motive force. An experimental study. Physiol. Plant. **2**, 157—163.
— 1950: A localized effect of light on the protoplasmic viscosity of plant cells. Nature **166**, 485.
— 1951: The effect of light on the protoplasmic viscosity. Physiol. Plant. **4**, 255—357.
— and L. Ehrenberg, 1953: Effects of β- and γ-rays on the protoplasmic viscosity of *Helodea* cells. Physiol. Plant. **6**, 159—165.
Votava, A., 1914: Beiträge zur Kenntnis der Inhaltskörper und der Membran der Characeen. Österr. bot. Z. **61**, 442—455.
Vouk, V., 1910: Untersuchungen über die Bewegungen der Plasmodien. I. Die Rhythmik der Protoplasmaströmung. S. B. kais. Akad. Wiss. Wien, math.-naturw. Kl. **119**, 853—876.
— 1913: Untersuchungen über die Bewegungen der Plasmodien. II. Studien über die Protoplasmaströmung. Denkschr. kais. Akad. Wiss. Wien, math.-naturw. Kl. **88**, 652—692.
Vries, H. de, 1885: Über die Bedeutung der Cirkulation und der Rotation des Protoplasma für den Stofftransport in der Pflanze. Bot. Ztg. **43**, 1—6, 17—26.
Wanner, H., 1945: Über die Wirkungen der Röntgenstrahlen auf das Plasma vegetativer Pflanzenzellen. Schweiz. Z. allg. Path. Bakt. **8** (Suppl.), 1—64.

WATANABE A., M. KODATI, und S. KINOSHITA, 1937: Über die Beziehung zwischen der Protoplasmaströmung und den elektrischen Potentialveränderungen bei Myxomyceten. (Japanese with German résumé.) Bot. Mag. (Tokyo) **51**, 337—349.
— — — 1938: Über die negative Galvanotaxis der Myxomyceten-Plasmodien. (Japanese with German résumé.) Bot. Mag. (Tokyo) **52**, 441—445.
— — — 1939 a: Über den Einfluß des Wuchsstoffes auf das bioelektrische Potential der Myxomyceten-Plasmodien. (Japanese with German résumé.) Bot. Mag. (Tokyo) **53**, 32—42.
— — — 1939 b: Über die Einflüsse des Wuchsstoffs und der anderen Substanzen auf die Potentialdifferenzen zwischen dem Plasmodium und dem Außenmedium. (Japanese with German résumé.) Bot. Mag. (Tokyo) **53**, 460—465.
WEBER, F., 1924: Methoden der Viskositätsbestimmung des lebenden Protoplasmas. Abderhaldens Handb. biol. Arbeitsmethoden **11**, Teil 2, 655—718.
— 1927 a: Review of BĚLEHRÁDEK's papers (1924/25, 1925). Protoplasma **1**, 298—300.
— 1927 b: Cytoplasma- und Kern-Zustandsänderungen bei Schließzellen. Protoplasma **2**, 305—311.
— 1929: Plasmolysezeit und Lichtwirkung. Protoplasma **7**, 256—258.
— 1930: Vakuolen-Kontraktion vital gefärbter *Elodea*-Zellen. Protoplasma **9**, 106—119.
WEBER, H. H., 1958: The motility of muscle and cells. Cambridge, Mass.
WEIDE, A., 1939: Beobachtungen an Plasmaexplantaten von *Phycomyces*. Arch. exper. Zellforsch. **23**, 299—337.
WENT, F. W., 1928: Wuchsstoff und Wachstum. Rec. trav. bot. Néerl. **25**, 1—116.
— 1938: Remarks about two auxin problems. Chronica Botanica **4**, 503—505.
WIGAND, A., 1885: Studien über die Protoplasmaströmung in der Pflanzenzelle. Bot. Hefte **1**, Nr. VI, 169—224.
WILLIAMS, M., 1923: Observations on the action of X-rays on plant cells. Ann. Bot. **37**, 217—223.
— 1925: Some observations on the action of radium on certain plant cells. Ann. Bot. **39**, 547—562.
WINER, B. J., and A. R. MOORE, 1941: Reactions of the plasmodium *Physarum polycephalum* to physico-chemical changes in the environment. Biodynamica 3, 323—345.
WORONIN, N., 1866: Beiträge zur Morphologie und Physiologie der Pilze. II. Reihe.
YAMAHA, G., und M. IDA, 1933 a: Protoplasmaströmung (Literatur). Protoplasma **17**, 146—158.
— — 1933 b: Untersuchungen der Protoplasmaströmung. I. Einteilung und Einfluß der Temperatur. (Japanese.) Shokubutsu oyobi Dobutsu (Botany and Zoology) 1, 461—474.
YOTSUYANAGI, Y., 1953 a: Recherches sur les phénomènes moteurs dans les fragments de protoplasme isolés. I. Mouvement rotatoire et le processus de son apparition. Cytologia **18**, 146—156.
— 1953 b: Recherches sur les phénomènes moteurs dans les fragment de protoplasme isolés. II. Mouvements divers déterminés par la condition de milieu. Cytologia **18**, 202—217.
ZURZYCKA, A., 1951: The influence of the wave length of light on the movements of chloroplasts in *Lemna trisulca* L. Acta Soc. Bot. Poloniae **21**, 17—37.
— and J. ZURZYCKI, 1950: The influence of temperature on phototactic movements of chloroplasts. Acta Soc. Bot. Poloniae **20**, 665—680.
— — 1957: Cinematographic studies on phototactic movements of chloroplasts. Acta Soc. Bot. Poloniae **26**, 177—206.
ZURZYCKI, J., 1951: The influence of temperature on the protoplasmic streaming in *Elodea densa* Casp. Acta Soc. Bot. Poloniae **21**, 241—264.
-- 1955: Chloroplasts arrangement as a factor in photosynthesis. Acta Soc. Bot. Poloniae **24**, 27—63.

Author Index

Subject Index

GPSR Compliance

The European Union's (EU) General Product Safety Regulation (GPSR) is a set of rules that requires consumer products to be safe and our obligations to ensure this.

If you have any concerns about our products, you can contact us on ProductSafety@springernature.com

In case Publisher is established outside the EU, the EU authorized representative is:

Springer Nature Customer Service Center GmbH
Europaplatz 3
69115 Heidelberg, Germany

Batch number: 10584556

Printed by Printforce, the Netherlands